3D Character Development Workshop

3D Character Development Workshop

Rigging Fundamentals for Artists and Animators

Erik Van Horn
University of the Arts

Mercury Learning and Information
Dulles, Virginia
Boston, Massachusetts
New Delhi

Publisher: David Pallai

Mercury Learning and Information
22841 Quicksilver Drive
Dulles, VA 20166
info@merclearning.com
www.merclearning.com
(800) 232-0223

Erik Van Horn. *3D Character Development Workshop: Rigging Fundamentals for Artists and Animators.*
ISBN: 978-1-683921-70-7

Library of Congress Control Number: 2017964298

181920321 This book is printed on acid-free paper in the United States of America

Our titles are available for adoption, license, or bulk purchase by institutions, corporations, etc. *Digital versions of this title are available at www.authorcloudware.com and most digital vendors. Companion disc files are available for downloading by contacting info@merclearning.com.* For additional information, please contact the Customer Service Dept. at (800) 232-0223 (toll free).

This book is dedicated to my wife, KB and our kids, Sage and McKenna.

Contents

Preface

When I worked in the training department at Disney Animation studios, our Character Technical Directors (hereafter, *character TDs*) worked in a department called Character Development. I liked the term *character development*, a more generalized term than the one most commonly used in the industry, *rigging*, as the whole process of bringing a CG animated character to life involves a lot of interrelated steps, of which rigging is but one. So when I was tasked to cross-train other departments in the esoteric art of character technical direction, I created the Character Development Workshop. The goal was to help the specialists in other disciplines of 3D animation to better understand the whole of the character pipeline and how it interfaces with their own concerns. Hence, the original audience for this set of tutorials were professional 3D animators, modelers, effects artists, texture painters, and shot finalers (lighter/compositors). Because character TD work is one of the most esoteric yet critical steps in the 3D animation pipeline, we felt it was important to shine a light on this invisible art.

Character development is a distinct art form, crucial to the success of a 3D production. In fact, I'd say no department in the pipeline has more of a direct impact on the perceived quality of 3D animation than character development, which includes rigging and skinning. Because riggers create the tools that 3D animators work with, the ultimate success of the performance begins with the quality of the rigs. But rigging cannot be understood in isolation; the development process involves a back-and-forth between character designers, modelers, and animators; it is the glue that holds all of these more "visible" art forms together.

So invisible is this art that I still run into clients and even animation distributors that don't know that this is a stage in the process that needs as much time and money budgeted to it as the stage before it (modeling) and the stage after it (animating). In over twenty years of independent contracting for the animation industry, I can't count the number of times I responded to a job ad for a "3D animator," received a creative brief with no mention of rigging, and then showed up for the first day on the job to find that none of the characters were rigged. Luckily, I just so happened to have had some rigging experience, otherwise I would have been ill equipped to handle these contract jobs, as it was apparent that the project owners had no idea that "animators" are not necessarily "riggers" (if they even knew rigging was a thing).

Adding to the confusion is the fact that all established companies use *auto-riggers*, or semi-automated rigging systems. When project owners hear about this, they envision that the rigging process is "taken care of" and they won't need to hire a character TD. What's often misunderstood is that auto-rigging systems were developed for consistency and to streamline redundant tasks, but they do not replace a qualified specialist, nor do they even guarantee saving time on a project. At best, they heighten the quality of the results of a project within the same time frame as manual rigging, by ensuring a baseline of consistency and usability. But they still require the guiding hand of a knowledgeable character TD to work as expected. Character TDs, in turn, need to know modeling, texturing/lighting, and animation to do their jobs correctly. This generalist background is not being taught in today's animation schools, in favor of a specialization training approach. The result is that good character TDs are rare and in high demand; at the same time, animators are crippled by not having the skills to rig their own characters, and are limited to royalty free rigs available on the Internet.

This set of tutorials was designed to fast-track understanding of the basic principles of 3D character development for a student or professional with some prior experience in 3D. The goal of this project is to grant animators and students the liberation of being able to design and develop original characters, no longer relying on technical assistance or preexisting rigs. Because this system is so fast and simple, it is designed for repeated use. Some former students have told me they continue to run through the process from time to time as a refresher course, and it always prepares them when they have to develop a new character for 3D animation.

My training method is built on three guiding principles:

- **The best way to learn a process is to do it**
- **Time is valuable**
- **The best way to understand one part of a process is to understand the whole process**

The first principle explains my **hands-on guided learning** methodology. I specify using Maya so that step-by-step processes can be explained; not that this makes it only useful for Maya artists, but since a free edition of Maya is available (and it is currently the industry standard tool) it creates a standard working environment so we can get specific with our step-by-step instructions. The fundamentals covered can be applied to any 3D software, but it is more intuitive to glean knowledge from a specific scenario than trying to parse vague abstract principles.

Which brings us to my second principle: **Time is valuable**. Nobody wants to sit through long-winded theory when they came to learn process. Since this training was developed for on-the-job professional animators, time is money. Theory can naturally be inferred through use-cases.

And the third principle is connected to the first two: **A holistic approach** is the only way to understand what happens when in the character pipeline. That is why we start with design, then modeling, then texturing, then rigging, then skinning, then animation, hitting the basics of each discipline; this is so a well-rounded character artist is aware of the entirety of the process no matter their specialization. For this reason, this workshop was mandatory training for many character artists at Disney Feature Animation.

Ultimately, the goal is to remove your fear of the technical hurdles, so you can feel empowered to start building your own rigs today and never feel like you need to rely on preexisting rigs so showcase your animation talents. You have stories to tell! They do not necessarily conform to rigs already available on the Web. Design your own characters, and tell your own stories! Once you fill in this crucial piece of the puzzle, the possibilities are limitless. Happy animating!

—Erik Van Horn

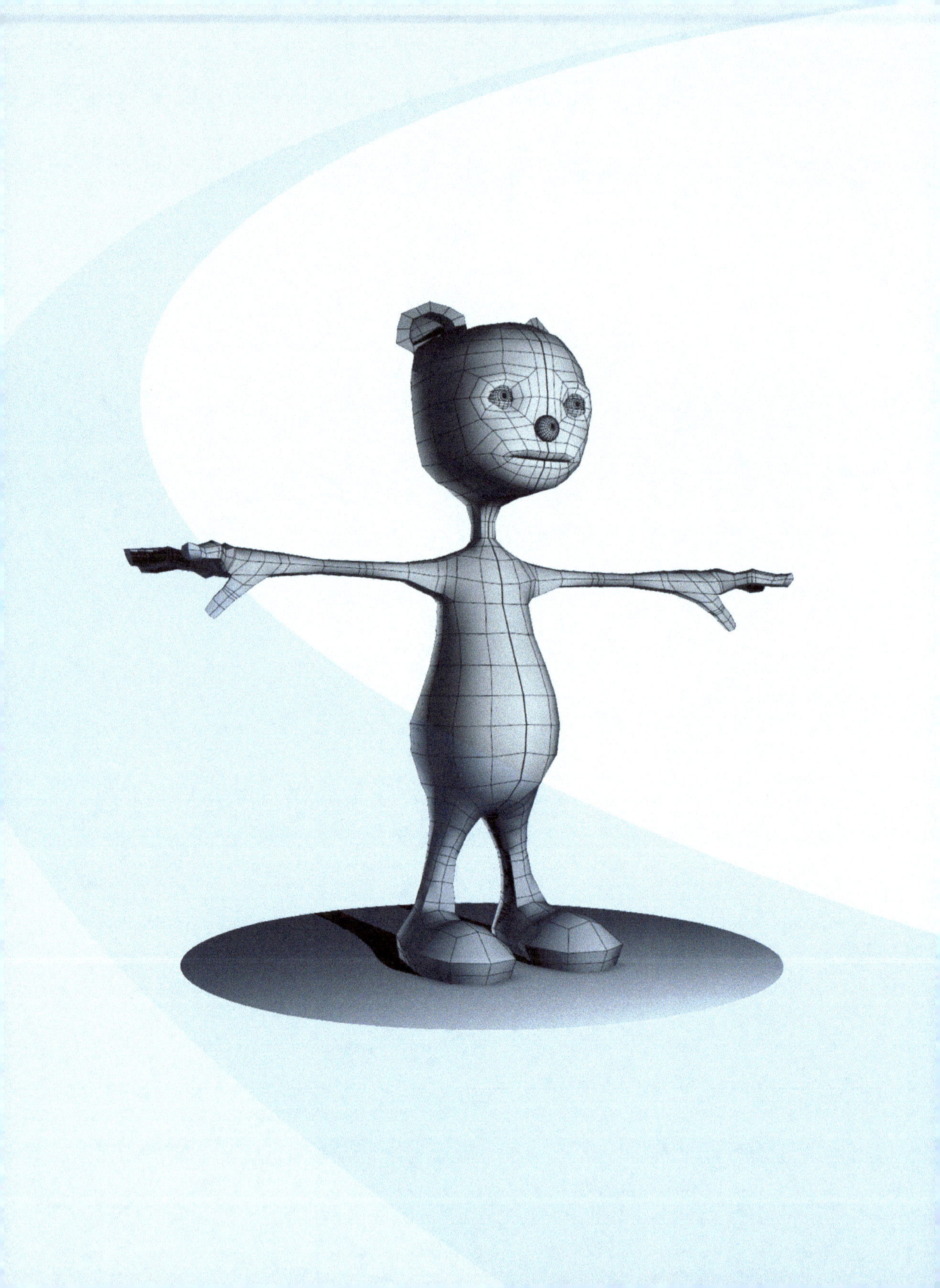

PART I
CHARACTER MODELING

1.1 INTRODUCTION AND OVERVIEW

This set of tutorials will empower you to rig your own characters. Whether you are a hobbyist, student, or professional animator, you need to have some understanding of the process of setting up a character for animation. Animation is a story medium. Characters drive story. 3D character development is the fulcrum of the character pipeline in 3D animation. Therefore, this is the most important and fundamental area of knowledge to the success of a 3D project.

However, it is also one of the most feared among students of animation and one of the most underappreciated among animation organizations. Why? Because it is a somewhat technical process within a creative field. Too often, animation studios draw a line between "technical directors" and "creatives" and organize their workforce into these two silos. This distinction is damaging, since it is essential for animators and other creatives in this field to be somewhat technical, and it is essential that technical directors—who are really technical *artists*—to be artists first, technicians second, because technical direction, development, and rigging are ultimately creative endeavors.

This series of tutorials will remove the fear you may have surrounding character setup. Step-by-step, we'll demystify the process by rigging an example character. A goofy example, at that, but the point isn't for this character to be a star in your next movie; the point is that this character will be your trial-and-error project; by the time you're done you'll have the confidence to design, model, and rig that character you *really* care about.

If step-by-step learning seems boring to you, you are in good company. I can think of many more interesting things to do, from skydiving to playing Tetris. But the purpose of this exercise is to run through the motions, so that these processes become second nature. The end result will be your ability to rig props and characters on-the-fly while animating, troubleshoot bad rigs that have been handed to you, or successfully articulate your desires and concerns to other character artists on your team. I promise, if you follow along and don't skip the steps, you will no longer feel lost or overwhelmed by the technical side of 3D character animation. Once you are able to model, rig, and animate, you will be a triple threat in the industry. Plus, each of your disciplines will inform and enhance the others; the best riggers are animators, and the best animators are riggers.

1.2 WHAT YOU NEED TO KNOW

The 3D Character Animation Workshop is meant to provide an overview of the character pipeline for students of animation or for specialists in any discipline of graphics or animation. This hands-on training assumes a basic understanding of 3D graphics and a familiarity with Maya's interface. If you don't have these, I recommend the "Getting Started" videos on the Maya Learning Channel (*https://www.youtube.com/c/mayahowtos*).

This tutorial uses Adobe Photoshop and Autodesk Maya for its examples. The main reason is that these are the industry standard 2D and 3D graphics editors, respectively. The secondary reason is that both are available for free download. Autodesk Maya offers a full-featured version available at Autodesk.com/education. You simply have to create an account to start the download. Photoshop's free full version is trial ware, but offers a lite, free version called Photoshop Express. You can also use Gimp (*https://www.gimp.org/*), paint.net (*https://www.getpaint.net/*) or any number of other free open-source alternatives.

If Maya isn't your main 3D package, no worries. The concepts and procedures are the same in any 3D package; only the terminology and interfaces are different. You'd be able to follow along with, say, Blender or 3ds Max, but it would be easier to follow along with Maya; all you need is some basic interface training to prep yourself. If you have 3D experience but no Maya experience, I recommend downloading the free version and running through the startup training on their website; it doesn't take much more than an afternoon, and you'd benefit from it in the long run anyway. If you spend any time in the freelance market, you are bound to run into Maya sooner or later, whether you're in games or film production.

The examples given are for the Windows versions of these software. If you are a Mac user, you'll still need a 3-button mouse to use Maya but you can hold the Control key while clicking to get the right mouse menus and use the command key where control-clicking would be done on Windows.

I also use some traditional abbreviations like *ctrl-click* for control-clicking and *LMB, MMB*, and *RMB* for left, middle, and right mouse buttons. Menu items and interface buttons are listed in bold with right arrows as separators, as per Maya Help Files tradition, like this: **Edit Mesh > Average Vertices**.

The version of Maya used in this edition was Maya 2018. Autodesk likes to move things around from release to release, so if you are using a later version you may discover that the menu item I refer to isn't in the specified path! If that happens, use **Help > Find Menu** to find the menu. That is, unless Autodesk moved **Find Menu** again.

Once you gather the required software and have a basic understanding of its interface, you'll be ready to move forward.

1.3 THE GENERIKAT TEMPLATE

The purpose of this tutorial is to provide an overview of the character development process in Maya using a very simplified character. We will use an archetypal "toon" style character. To jumpstart the process and get us on the same page, I've provided a Photoshop template file to enable you to customize your creation while following the basic process.

Open the layered Photoshop file included with the lesson files, "*generikat.psd*". In this file, you will find the layers as they appear in this screencap:

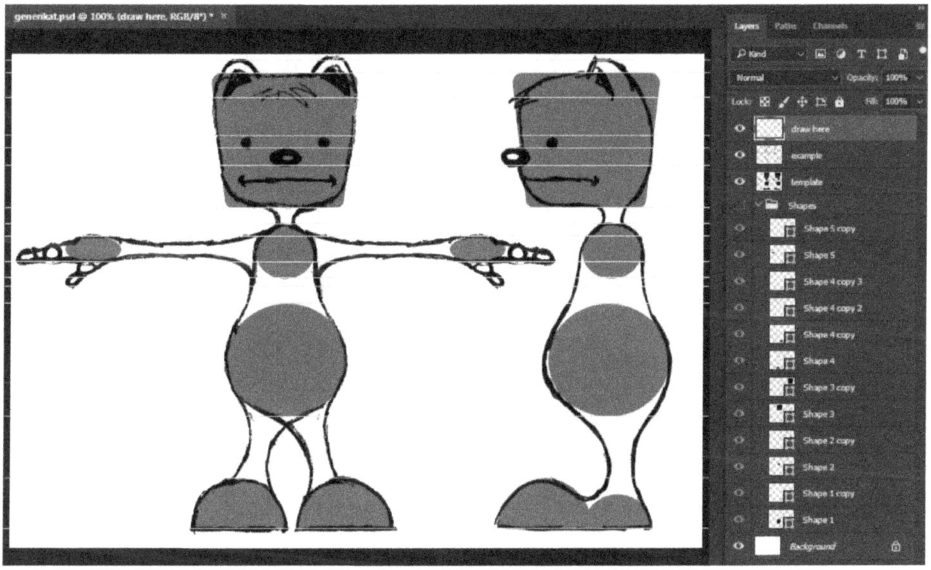

On the *example* layer, you'll find an ambiguous bear/cat-like design that you can use if you like. Or, hide the example layer and draw your own variation on the *draw here* layer. The shaded shape areas can be used as your guides: simply trace around the shapes and bridge them to create a unique character. Or, hide the *template* layer and show the *Shapes* layer group. These are the vector shapes from which the template layer was originally rasterized. By selecting layers with the Move tool (turn on **Auto Select** and **Show Transform Controls** on the options bar) you can manipulate their size and placement to create a new template for toon designs of different proportions.

TIP

Copy your line art layer, create a new channel and paste it in, and then **Image > Adjustments > Invert** (Ctrl+I) to create a negative of the line art in an alpha channel. Save the file as a 32-bit image (such as TIF or TGA) and the alpha channel will read as transparency in Maya. This means your image plane will only show the line art, not the white of the page. If you want, use the version I've already done this to: *generikat.tif* (available with the lesson files). Another option is to turn off all layers but the line art and save as a PNG file. Transparent pixels in a PNG will be honored by Maya.

1.4 IMPORTING AN IMAGE PLANE

Once you have something you are happy with, turn off all layers but the "draw here" layer and save it as a flat file. Put a copy in your Maya project directory in the ***sourceimages*** folder.

TIP

If you don't know where your Maya project directory is, open Maya and look at the path under **File > Project Window**. Unless you've changed it, it should be something like **C:/Users/[*username*]/Documents/maya/projects/default**

Bring it into Maya as an image plane in both the front and side views. (On the pane menubar, **View > Image Plane > Import Image**.) In the image plane attributes (Ctrl+A to toggle open the Attribute Editor), under **Placement Extras**, manipulate the **Image Center** values until the character is centered over the vertical axis in both viewss and standing flat-footed on the ground plane. (If you want to work larger you can also adjust the **Width** and **Height** values.)

Copy the **Width**, **Height**, and **Image Center Y** values between the two image planes to ensure the character lines up in both views.

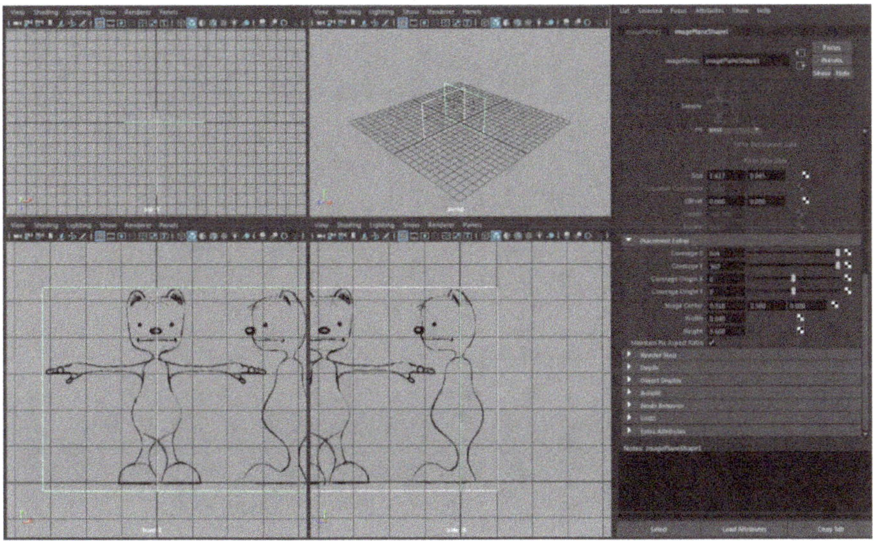

1.5 SETTING UP SYMMETRY

Create > Polygon Primitives > Cube. Move and scale it to just fit within the character's largest bulk; in this case, the belly. Give it a center line in all directions. In other words, under **INPUTS > polyCube1** in the channel box, **Subdivisions** set to "2" in all dimensions.

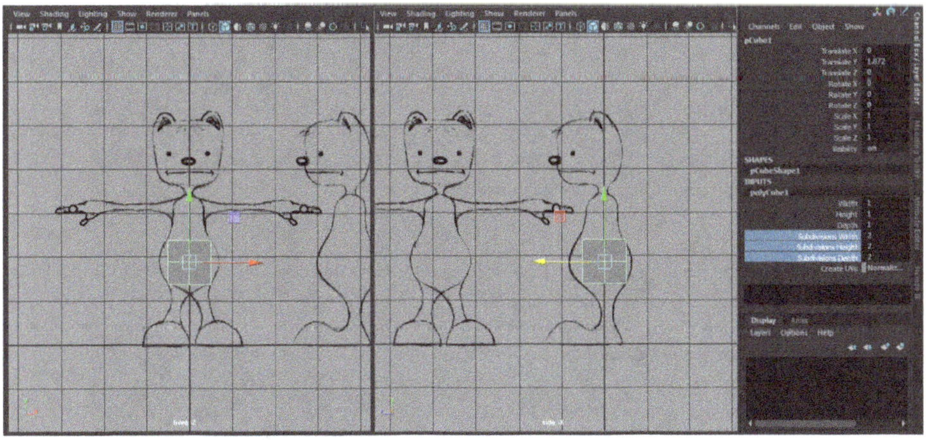

Go into face mode and delete half (all faces on one side of the y axis in the front view).

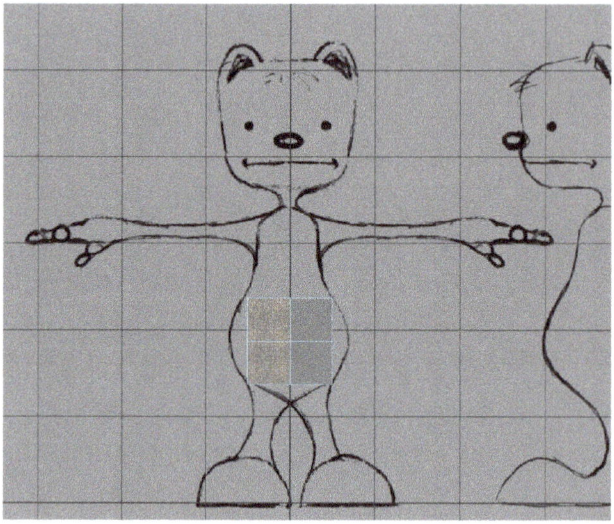

Go back to object mode (**F8**) and **Edit** > **Duplicate Special** > **Options**. Set it to "Instance" and scale negative 1 on the X. This will create a mirror across the X axis for symmetrical modeling. We will create the rest of the model with a series of extrusions; just remember from time to time to delete the inner faces that will be created anytime you extrude along the center line.

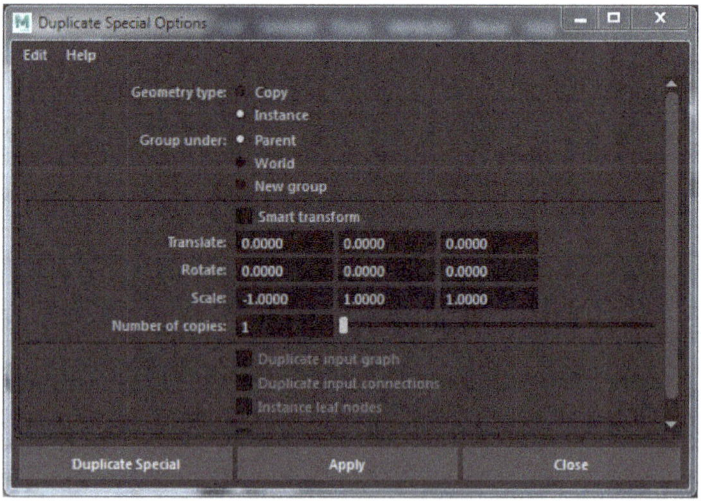

Select the top faces (of either instance) and **Edit Mesh > Extrude** (Ctrl+E) with 3 divisions, to the top of the torso.

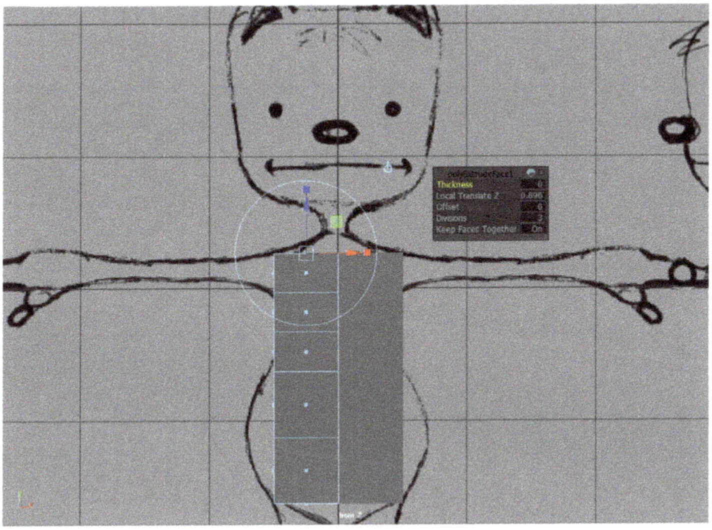

TIP

Window > Settings/Preferences > Preferences > Selection > Polygon selection: Select faces with Center. Use this setting in order to drag-select edge faces in the orthographic views more easily.

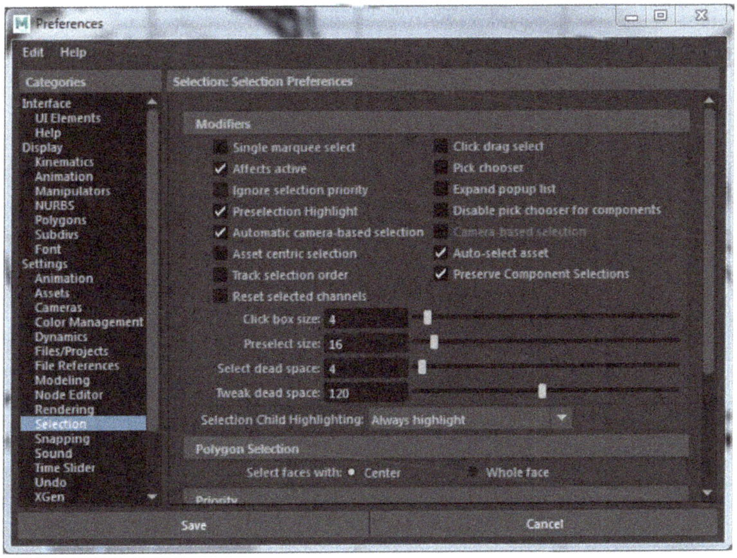

Remember to delete these inner faces and all faces along the centerline going forward. Hit the **3** key to smooth preview and it will remind you if you have hidden faces to delete. Hit **1** to return to unsmoothed mode.

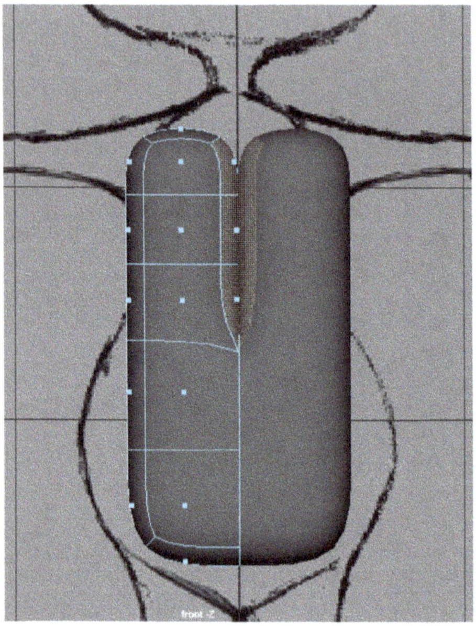

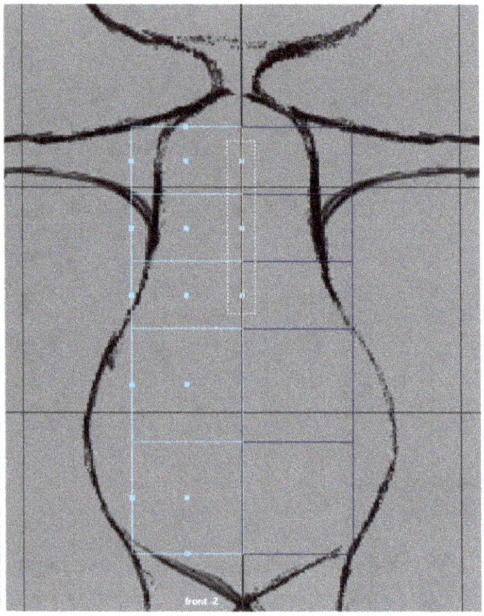

TIP

Double-click the Move Tool icon to get at its settings. Under **Symmetry Settings** turn on **Object X** and **Preserve Seam** to keep you from accidentally moving vertices from the center line. In case a center vertex ever does get moved, hold **X** to grid-snap it back to the center.

1.6 BLOCKING OUT THE BODY

In the front view, move the top corner verts to the outside edge of the neck, as in the following picture. Move the next pair of vertices below them to the top of the shoulder. (Okay, Generikat doesn't have clearly defined shoulders but place the vertices where you imagine the arm will rotate from.) Move the next pair below to the armpit.

TIP

Remember to drag-select rather than pick-select in the ortho views to get the unseen vertices behind. If you work in **Smooth Shade All** mode (hotkey **5**) you may have to adjust your Selection tool settings: under **Common Selection Options**, turn off **Automatic camera-based selection**. If this is on, smooth shaded viewports will not allow you to marquee-select components occluded from view. The other option is to switch to wireframe to marquee-select.

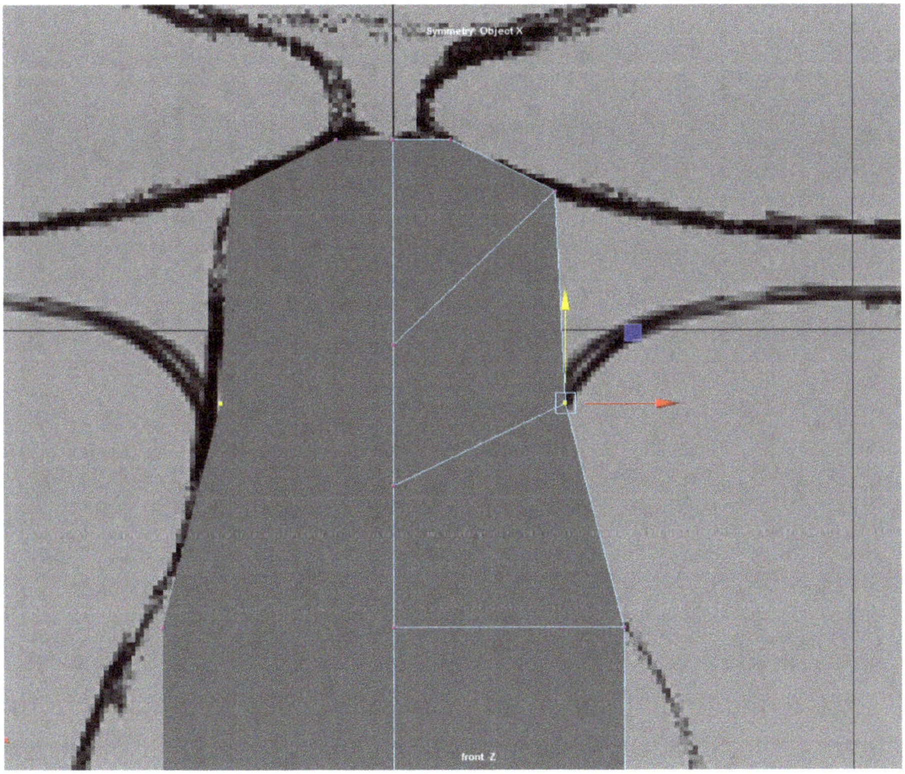

Move the bottommost side verts inward toward the crotch to represent the inside of the leg. Move the next vert pair above them down to become the outside of the leg. It should now look like this:

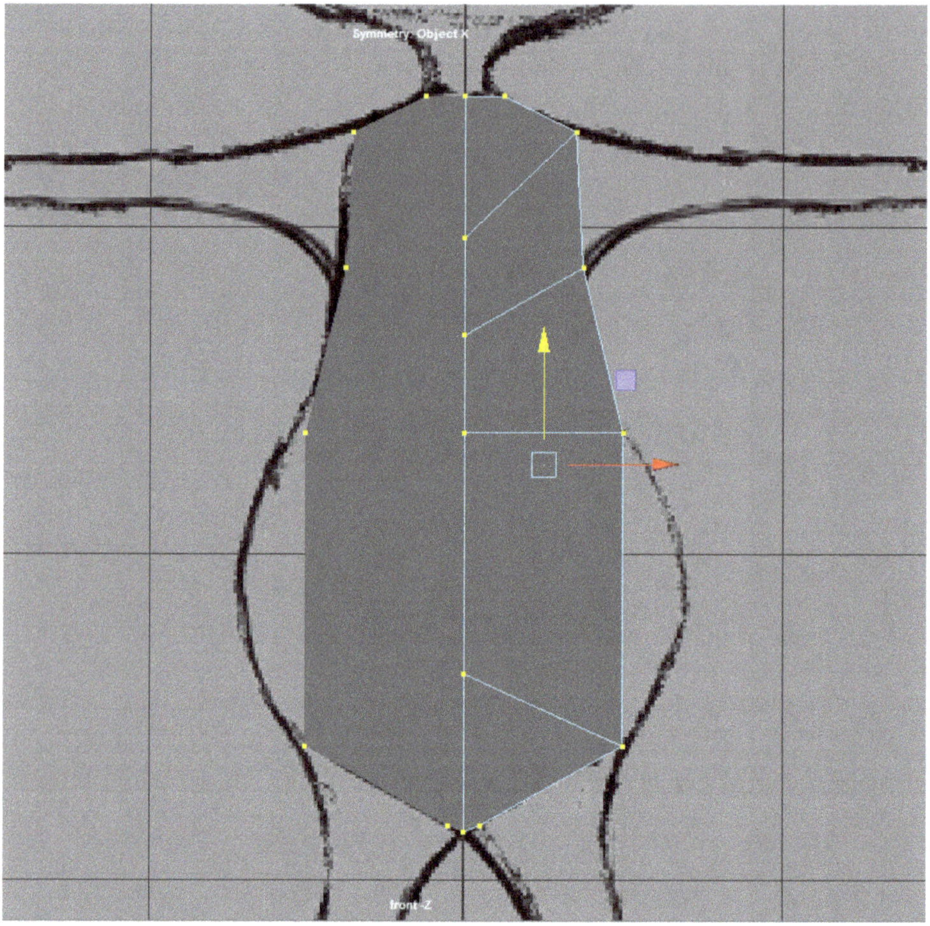

Notice the three bottom pairs of vertices at the crotch. It is important that extruded limbs and appendages have at least one edgeloop between them. By pinching in the outer edges toward the center line we've created a buffer between the center line and the faces we intend to extrude to get the legs. Remember this trick for all future characters.

Add a new edgeloop in the center of the belly. You can use the **Mesh Tools > Insert Edge Loop** tool, or I suggest a shelf button with the following MEL command: ***PolyConvertToRingAndSplit;***

This command will split any edge you currently have selected with a new edgeloop. Move the new vert to the character's silhouette.

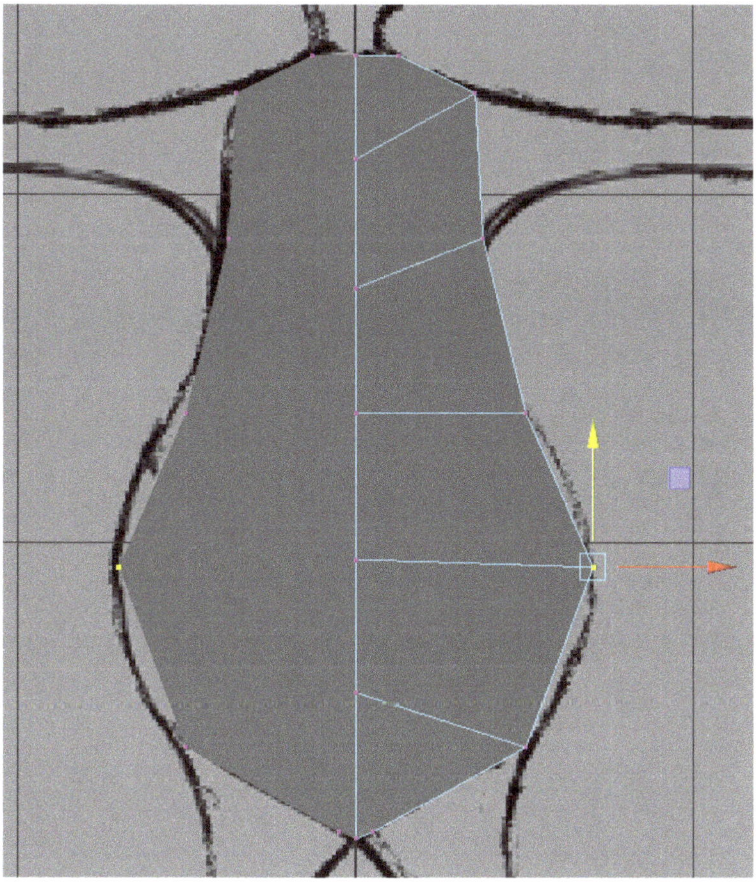

The rest of the process will follow this basic procedure: extrude to get new geometry, move verts to refine shape, add edge loops to further refine.

Next extrude for the arm. Select the arm face, extrude to the wrist with 4 divisions. The middle one will be the elbow.

TIP

You may want to turn OFF symmetry in the Move tool settings before extruding. Extruding in symmetry mode will cause your Extrude Manipulator tool to function strangely.

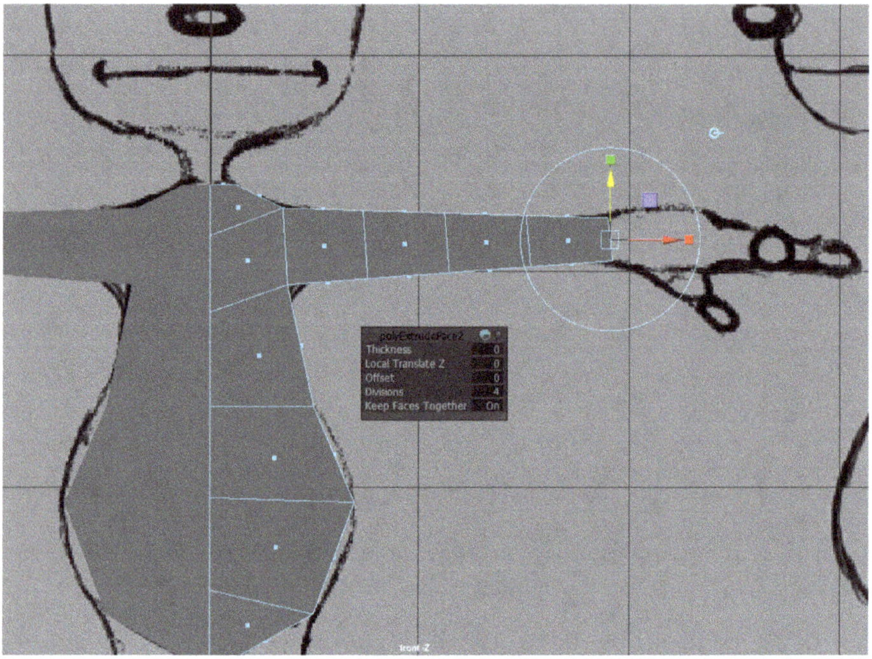

Move the verts to position using the image plane drawing as your guide.

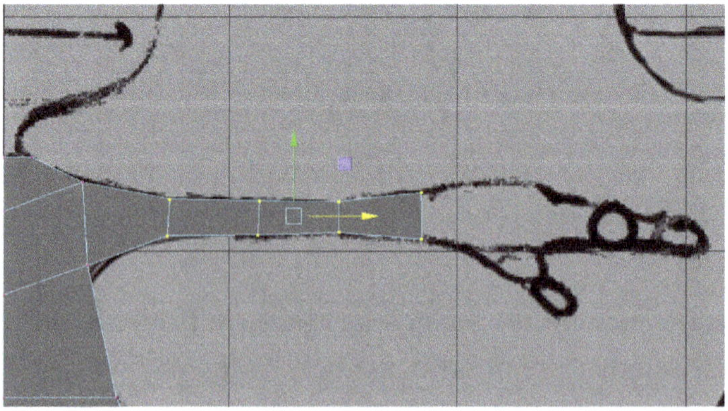

Do the same for the leg: select the face, extrude to ankle, divisions=4, move verts into place. The middle division should roughly be where you want the knee to bend.

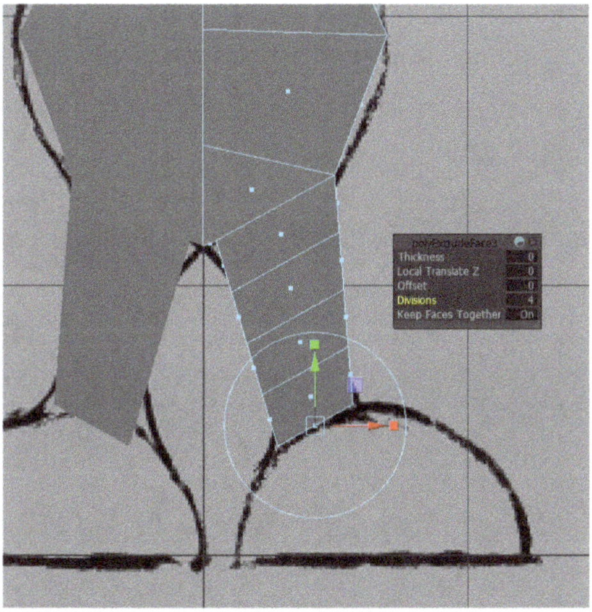

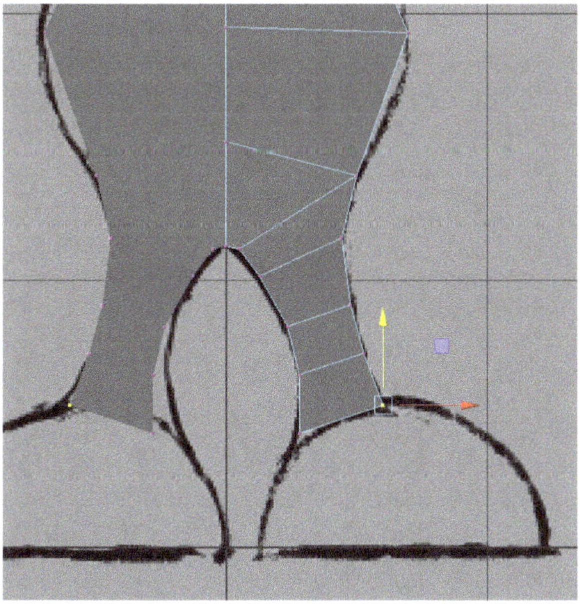

It's worth noting that riggers prefer things "on the grid" —that is, limbs and edgeloops arranged in vertical and horizontal lines as much as possible. However, here we have to splay Generikat's legs to keep his clodhopper feet apart.

Uh oh. Looks like we've been neglecting the side view. Move all the verts horizontally to the nearest character silhouette. Keep the center line in the center.

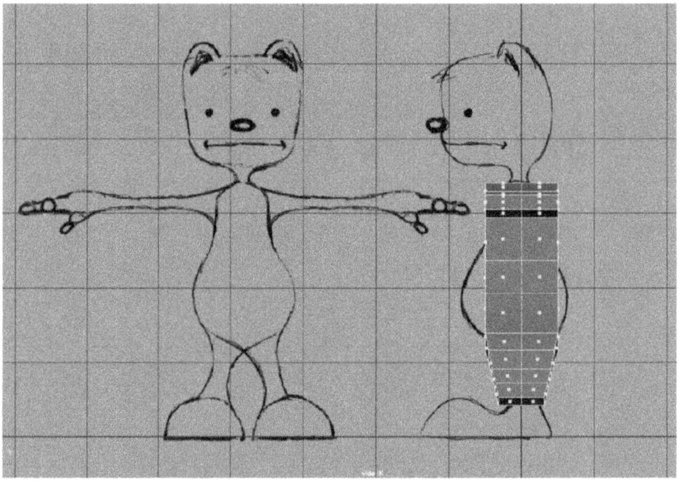

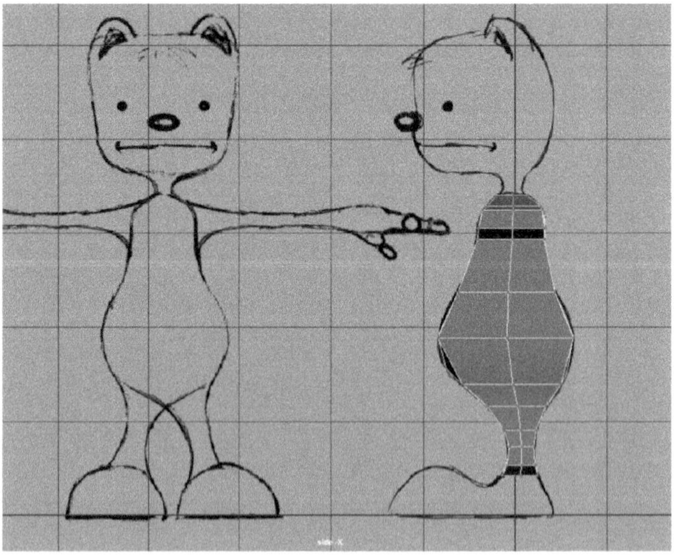

Now your model should look something like this:

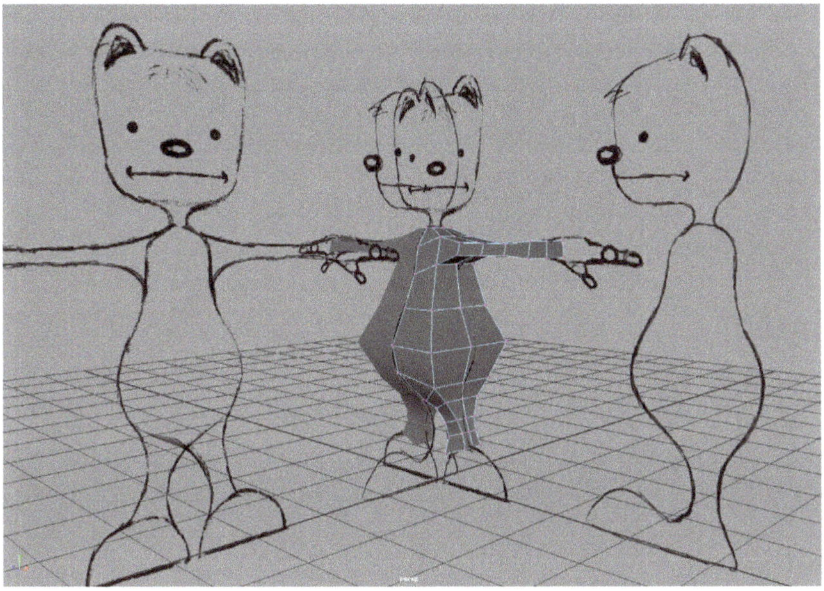

Next extrude the bottom leg-face down for the foot. Flatten out the verts so they lay flat on the ground plane. Add a division.

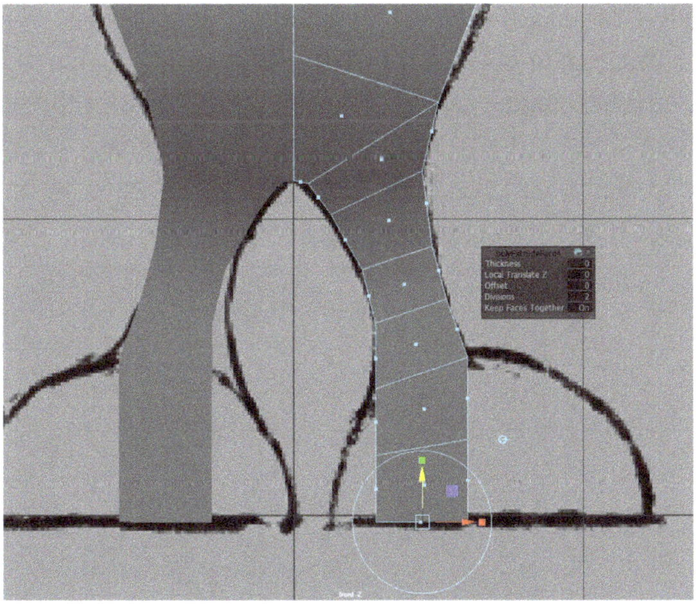

Next select the front faces of the last extrusion and extrude them forward in the side view to form the foot. Give it three divisions.

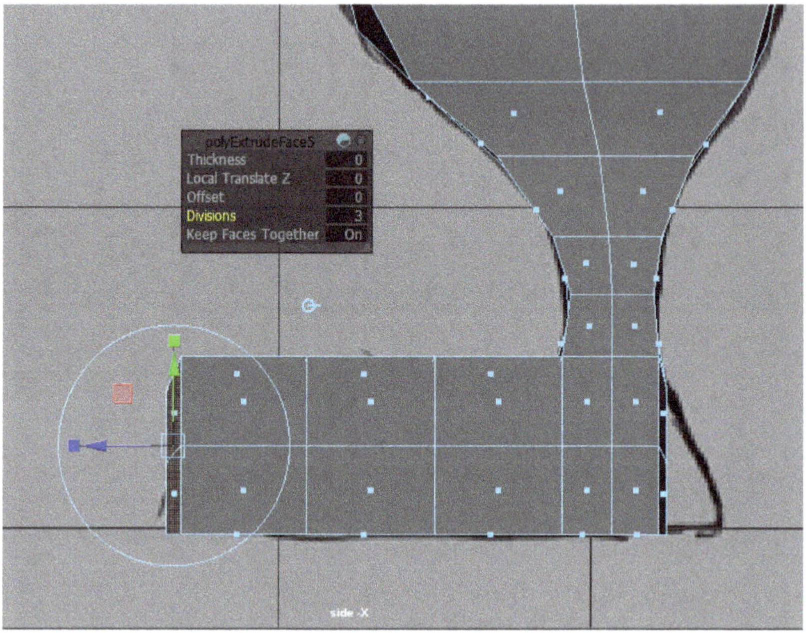

Fashion all these new vertices into the foot shape by dragging them in the side view. Feel free to use wireframe view or x-ray mode to better see your image plane. (**Panel menu > Shading > X-Ray.**) Man, those clodhoppers are huge. Now's the time to deviate from the template and make them smaller if you want to animate more easily later on. In fact, this portly character with huge feet is the worst-case scenario for animating; but hey, we're learning, right?

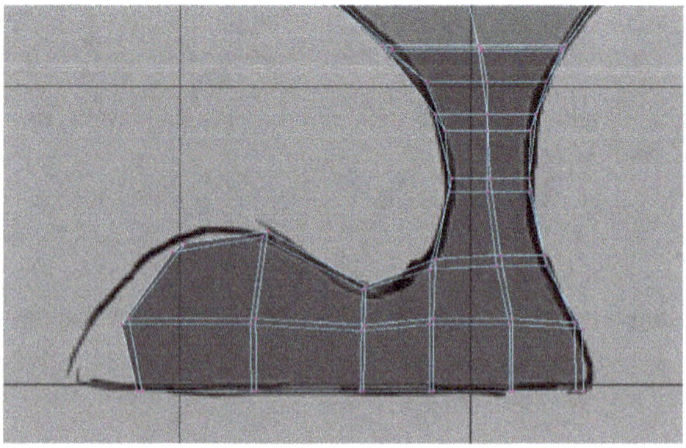

Switch to the front view and extrude the end-of-arm faces for the hand—pull it out to the end of the palm (beginning of fingers—approximately half the length of the overall hand). Give it three divisions.

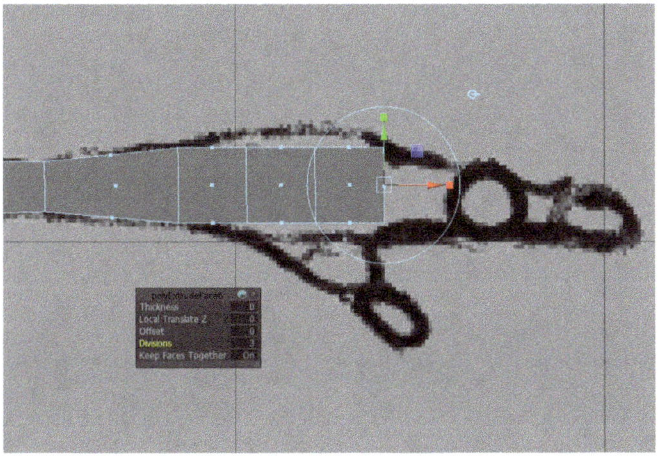

In the perspective view, grab the faces at the end and switch to the Scale tool (**R**). Scale them inward along Z until the middle two faces combined are the same width as each outer face. These will form three finger bases. The middle finger is currently split into two faces; we'll later split the others to match it. Since this is a toon-style character, we will be sticking with the cartoon convention of three fingers.

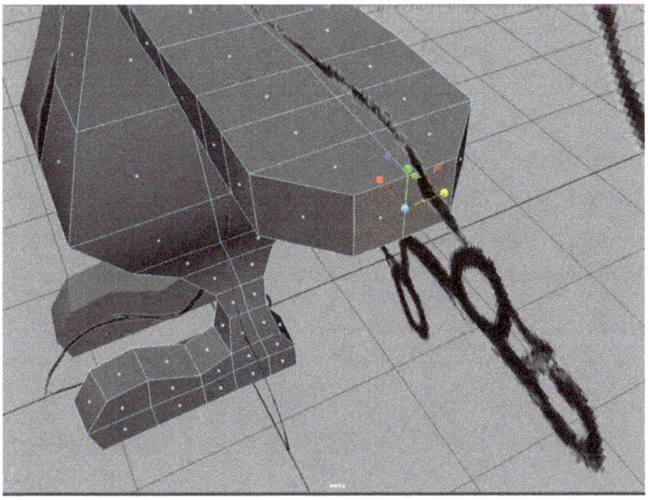

Select the first face and extrude for the index finger. Add divisions=4.

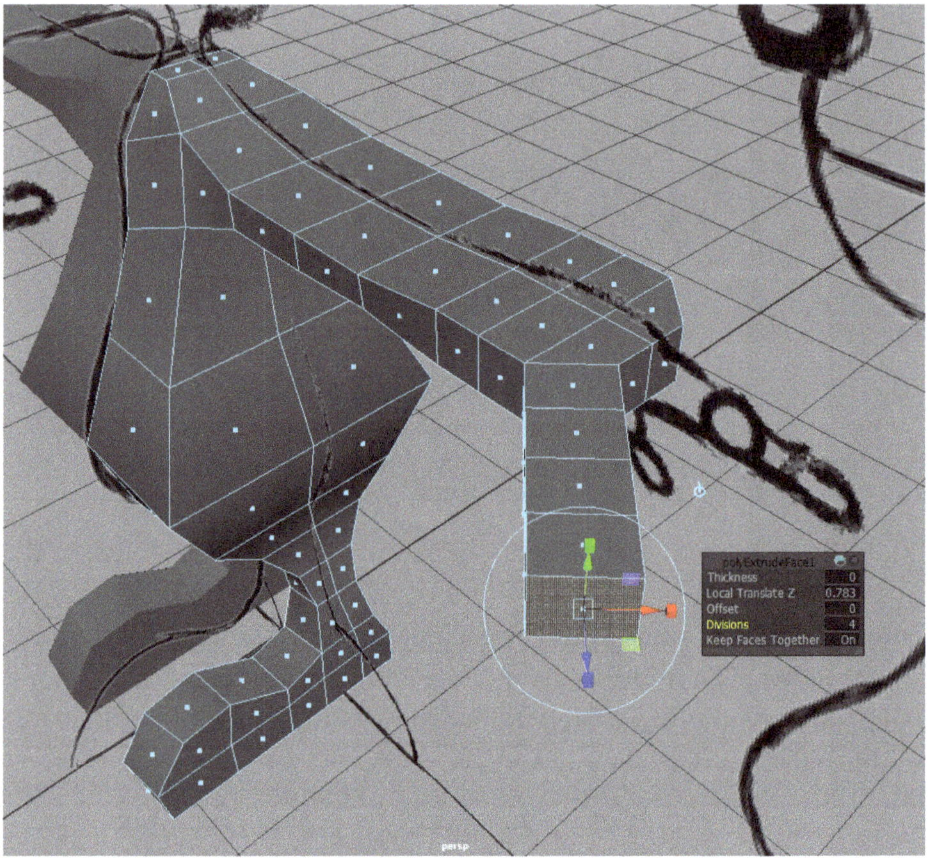

TIP

Remember you can work on either side of an instanced symmetrical model, but if you are working on the "scale x:-1" instance, your Manipulator tool will behave backwards.

Select the middle two coplanar faces and extrude the same distance for the middle finger. Divisions=4.

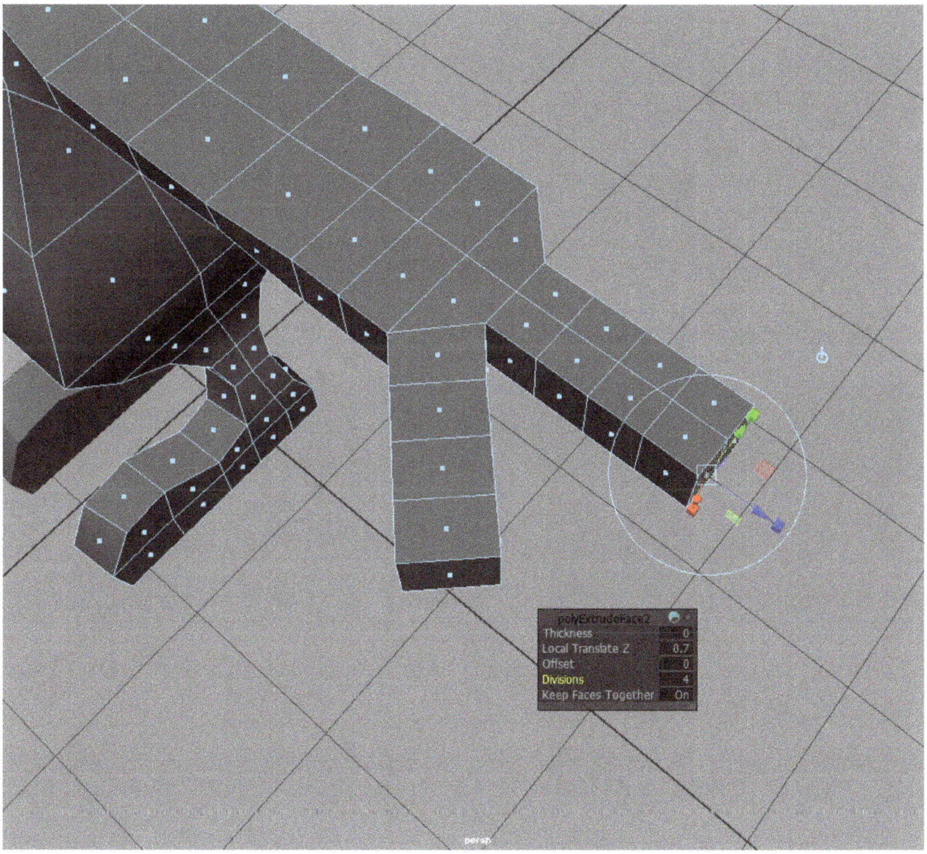

TIP

If the image planes get in your way, hide them (**Panel menu > Show** > uncheck **Image Planes**) or just give the image plane's **Image Center X/Z** a negative number to move it away from the model in the perspective view—it'll still appear the same size in the ortho views.

Then do the final finger.

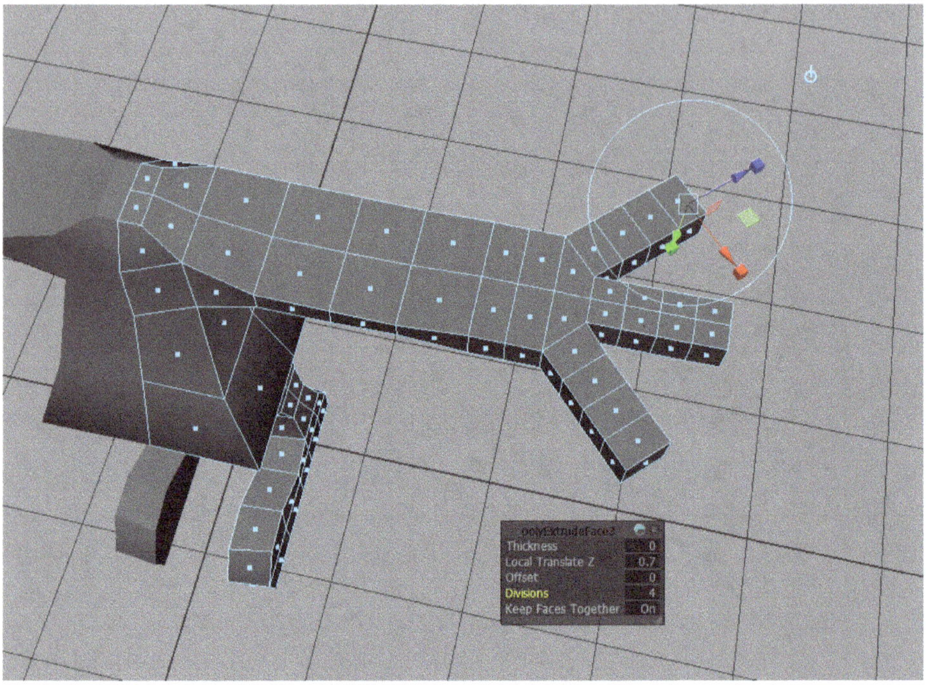

Now, swing to the bottom and select and extrude from this face to make the thumb:

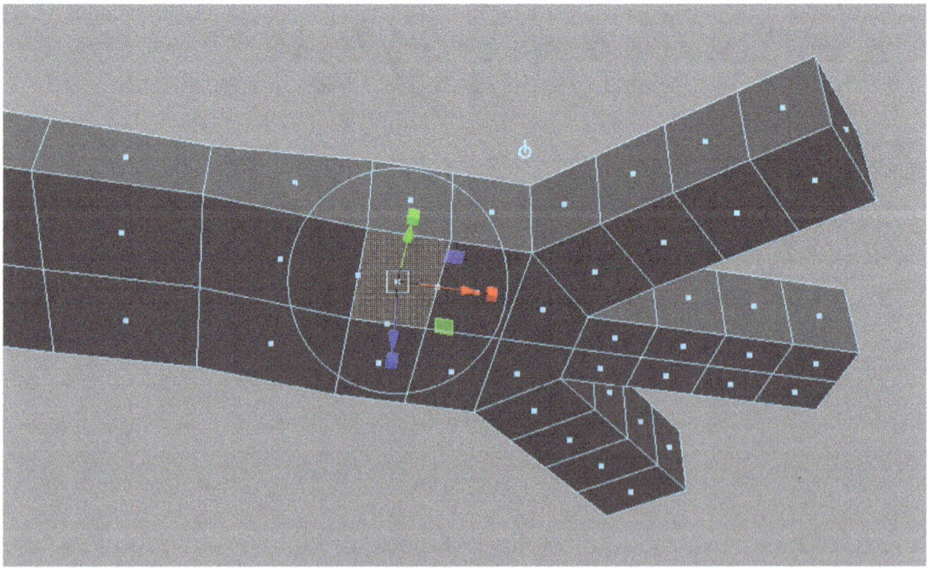

Switch to the front view and move verts to the thumb silhouette. It should protrude from the bottom of the palm at a 45-degree angle forward:

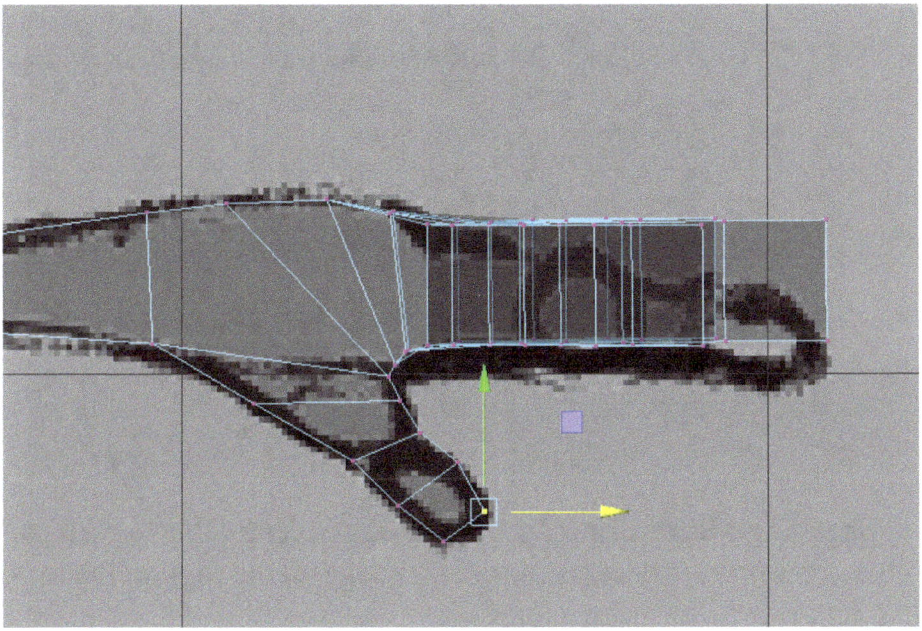

Now swing back to the top of the hand. Split the ring from the pinky to the index to create a center line on each digit, as shown:

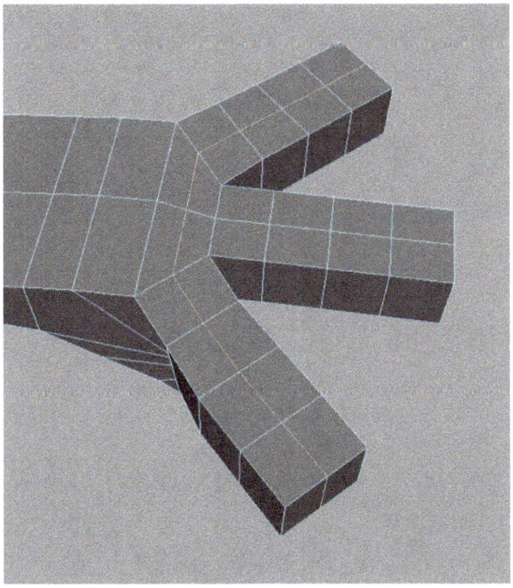

And split the thumb ring as well. Now all digits should have a center line running through them.

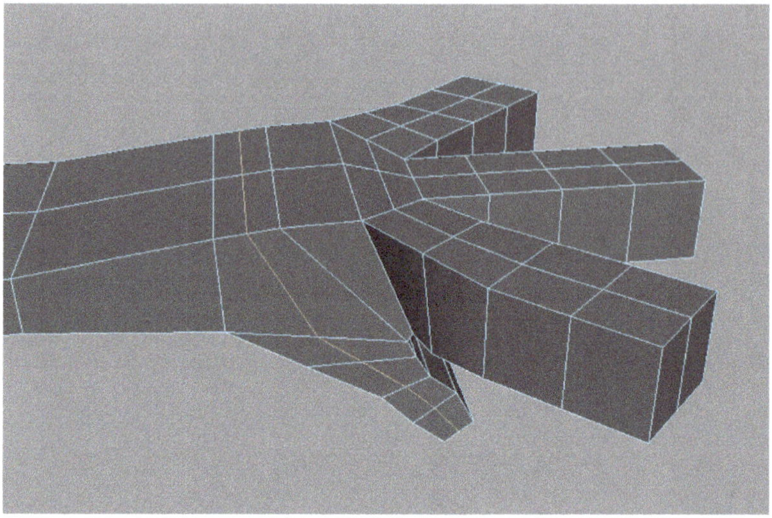

Speaking of center lines, it's time to add a center line to each half of our symmetry. Select any horizontal edge on the front of the torso and split them with an edgeloop:

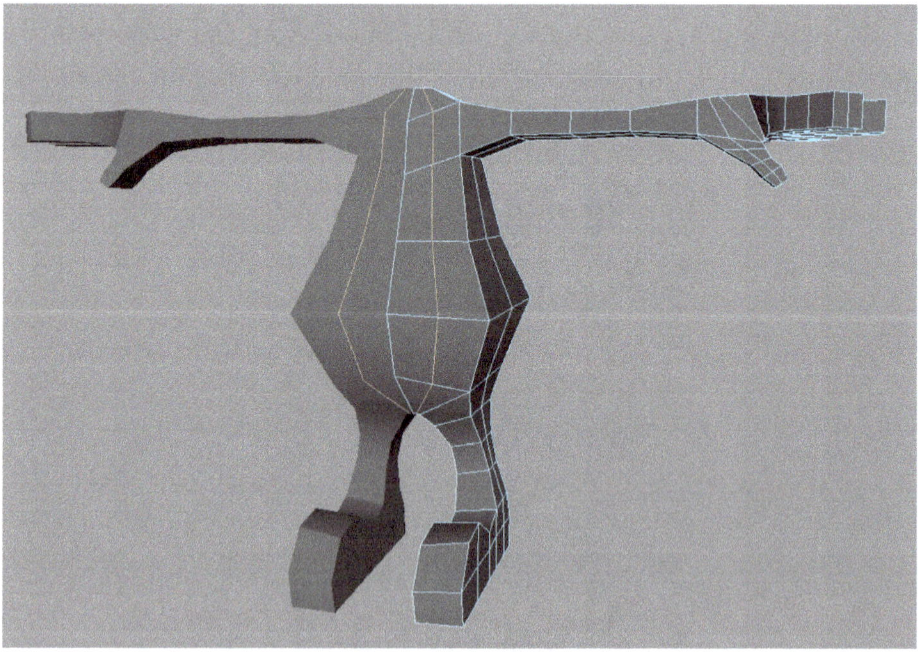

Do the same for the arm:

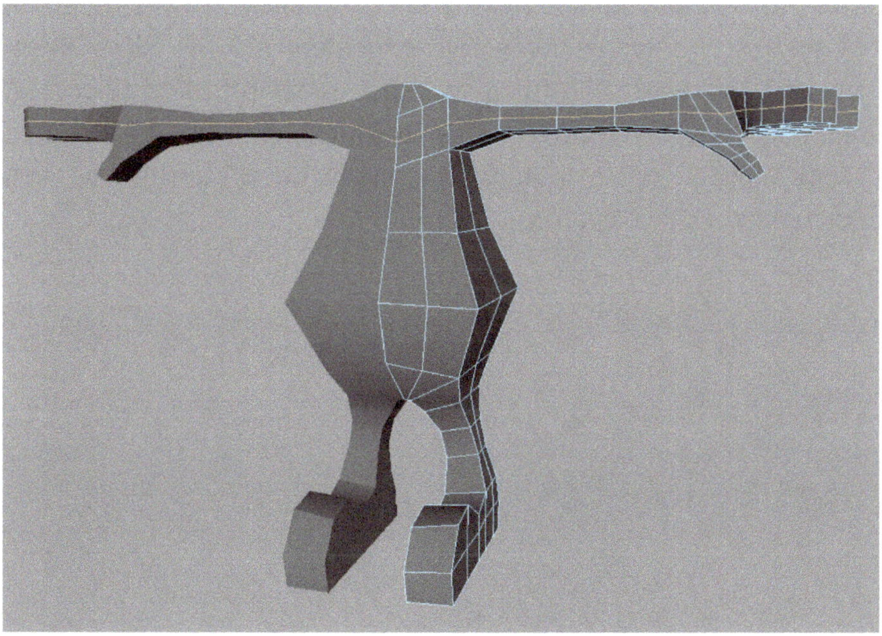

And leg:

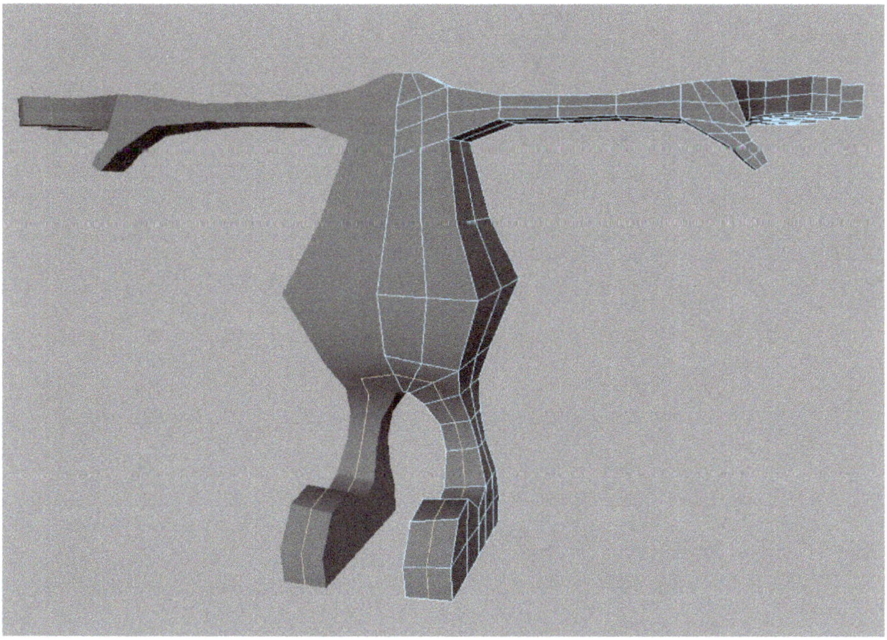

Now you should have enough vertices you can "punch back" the corners to create a less boxy shape. The fastest way is to double-click an edge to select the whole edgeloop and hit **Edit Mesh > Average Vertices** (make this a shelf button for repeated selecting of components and averaging to smooth the shape). Do this for every 90-degree edge. In this image, I've averaged the highlighted corner edgeloops two times apiece:

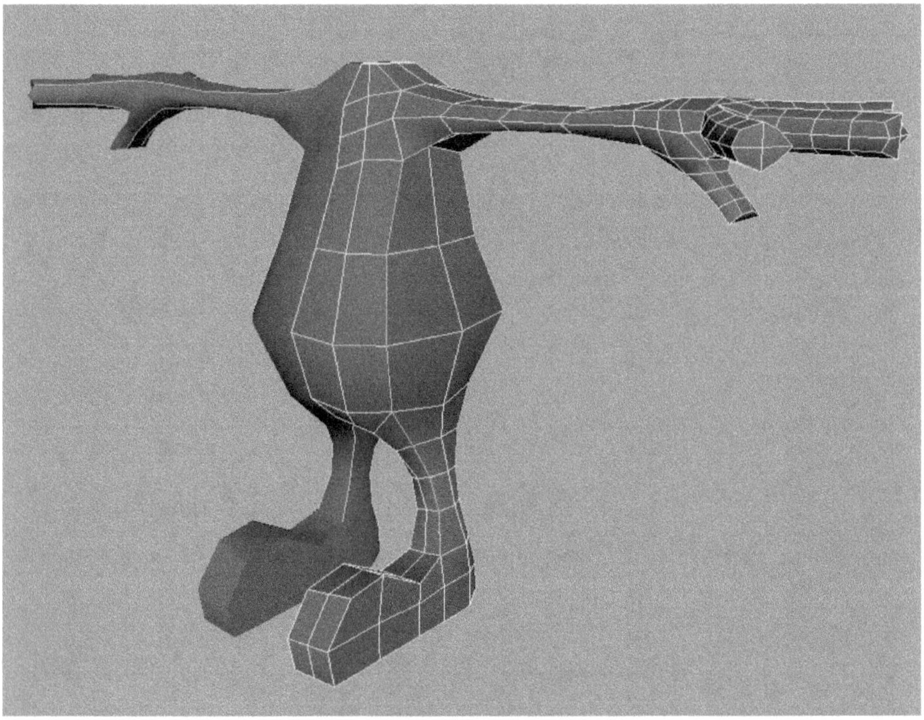

Also take a moment to thin out the arm from the top view.

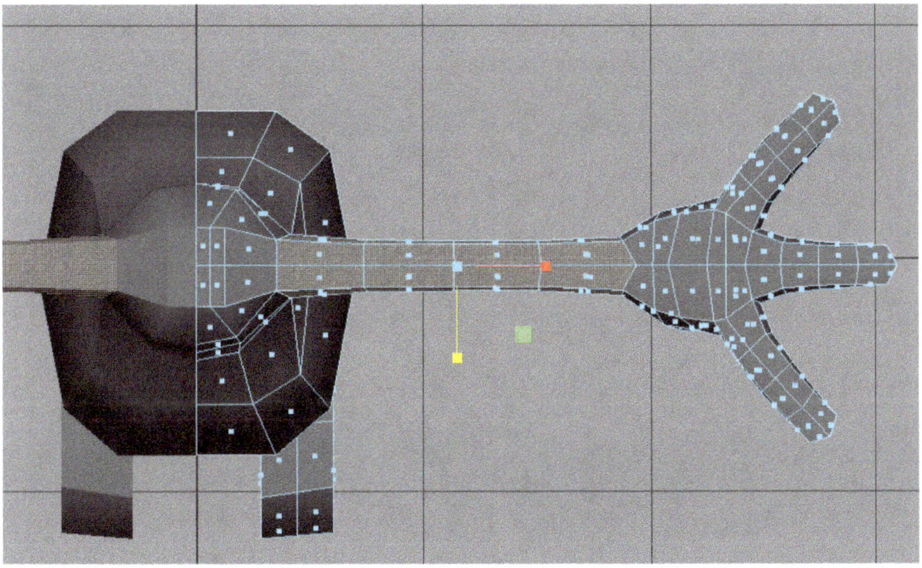

And tweak the foot vertices into place in the front and perspective view.

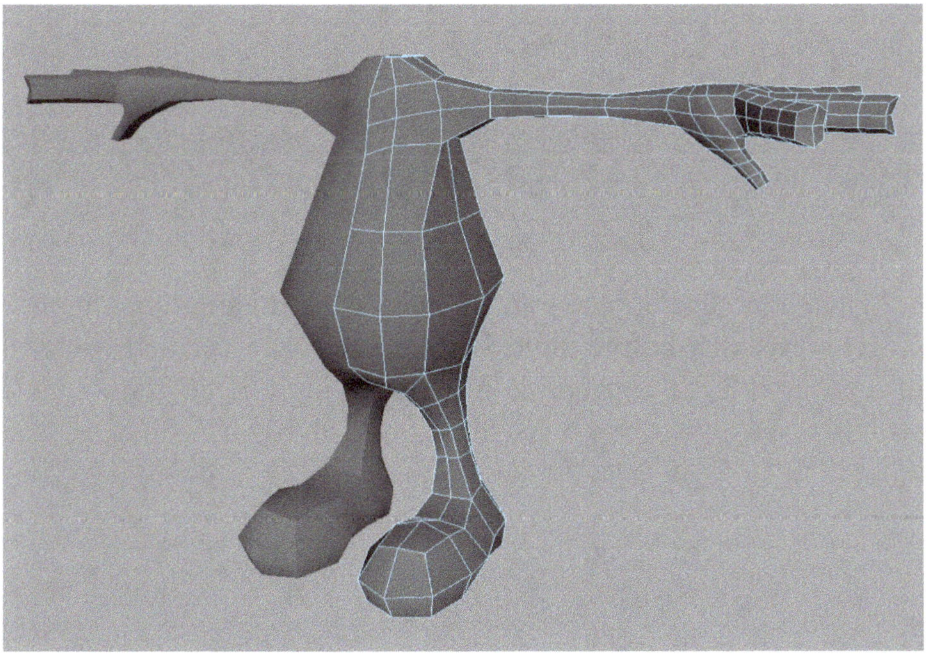

In the top view, double-click to select the edge rings between joints in each finger and scale them down.

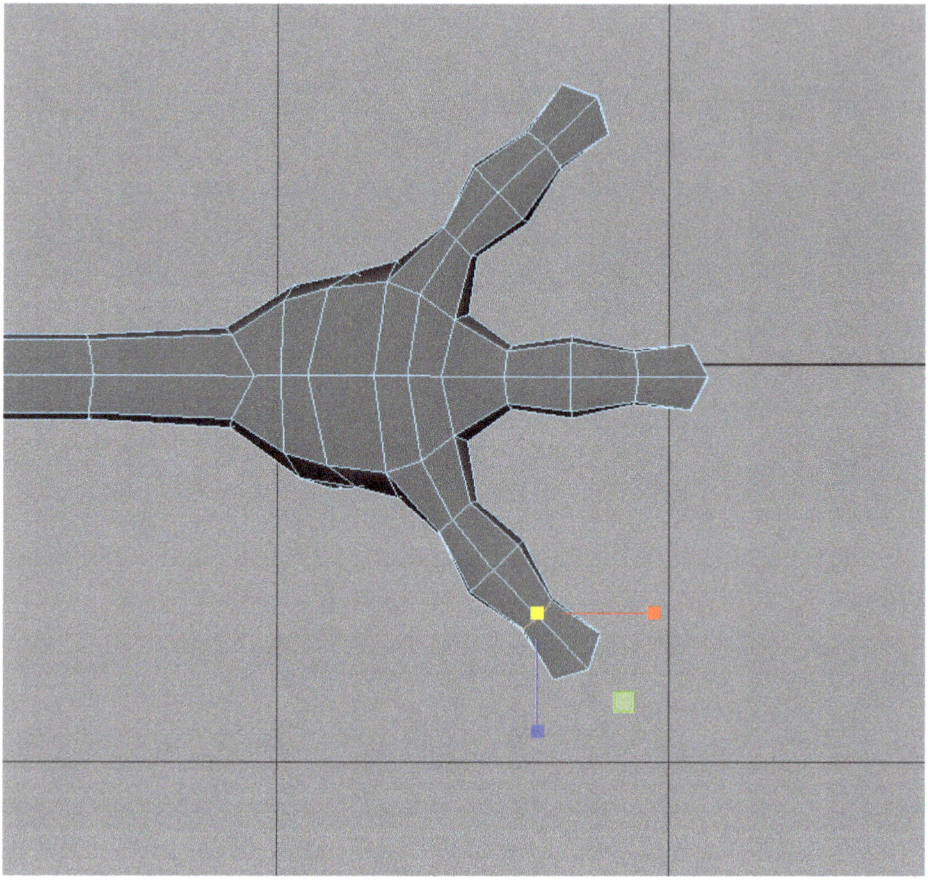

Then add edgeloops on either side of the middle loop on each finger—this will be the knuckle. Since this is a cartoon, we will also stick with the convention of only having one central knuckle per finger (leaving off the smaller outer knuckle). Also add an edgeloop close the base of each finger.

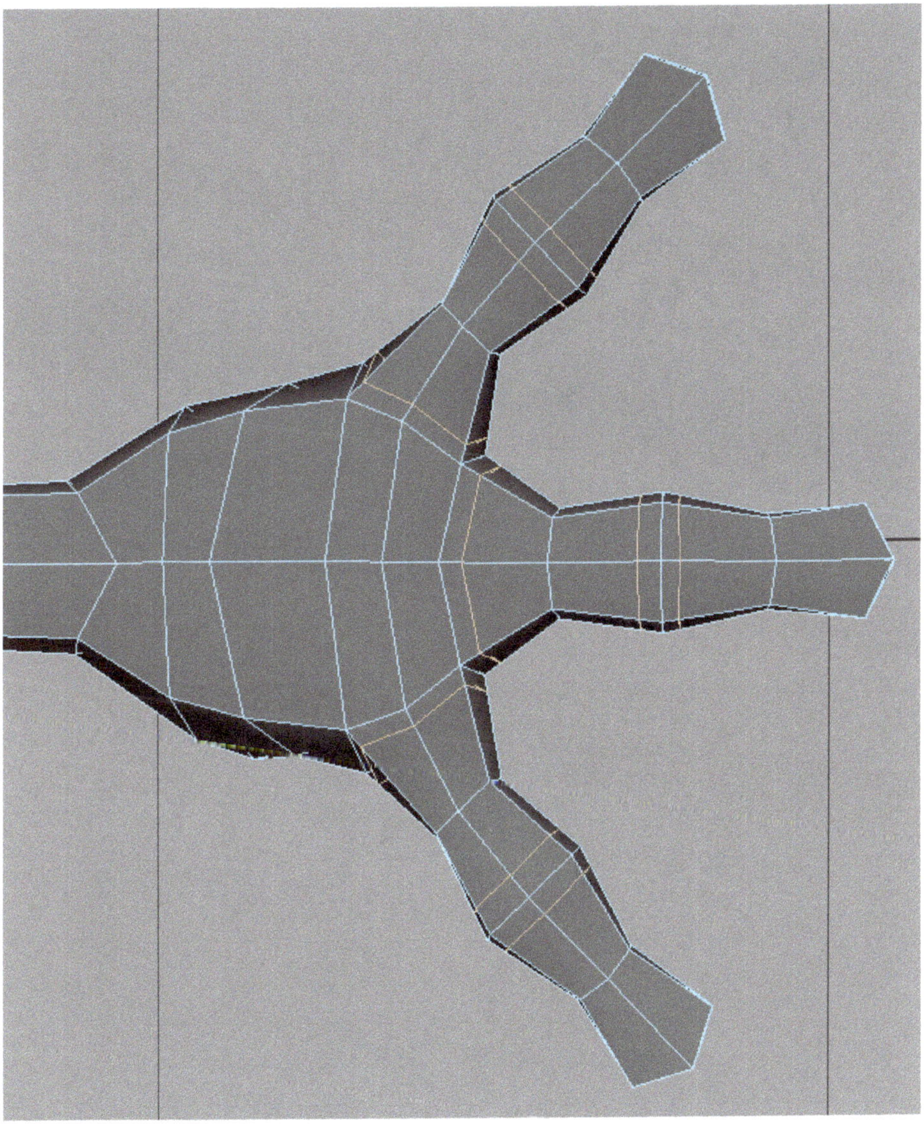

Also add an extra edgeloop at the wrist, either side of the elbow, and the shoulder.

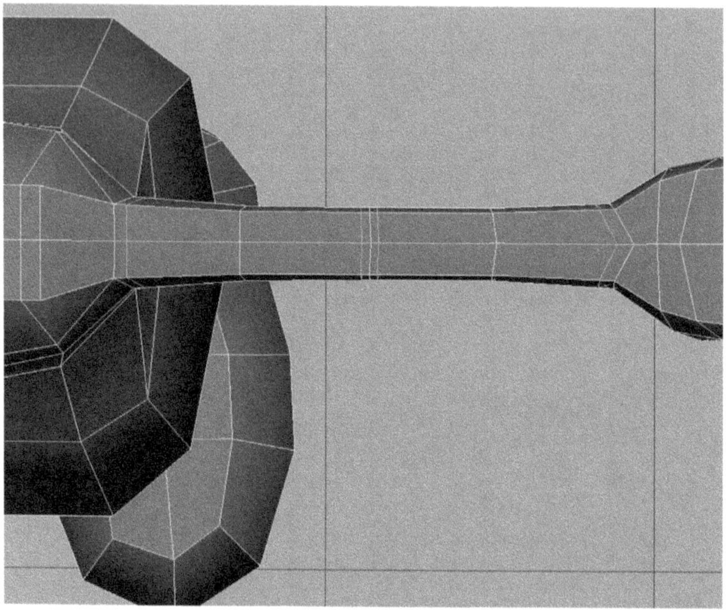

Do this for the legs too.

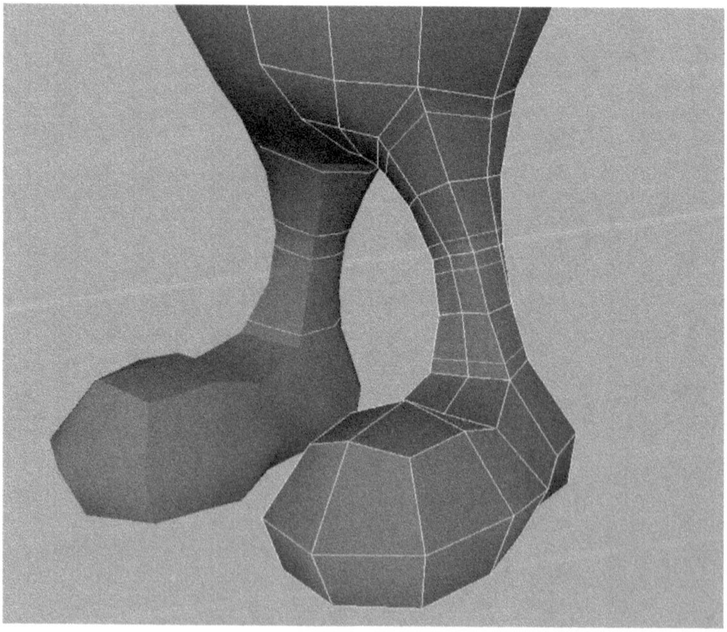

Shape the fingers if you have to, then check your work in smooth mesh preview mode (**3**). It is starting to look tooney!

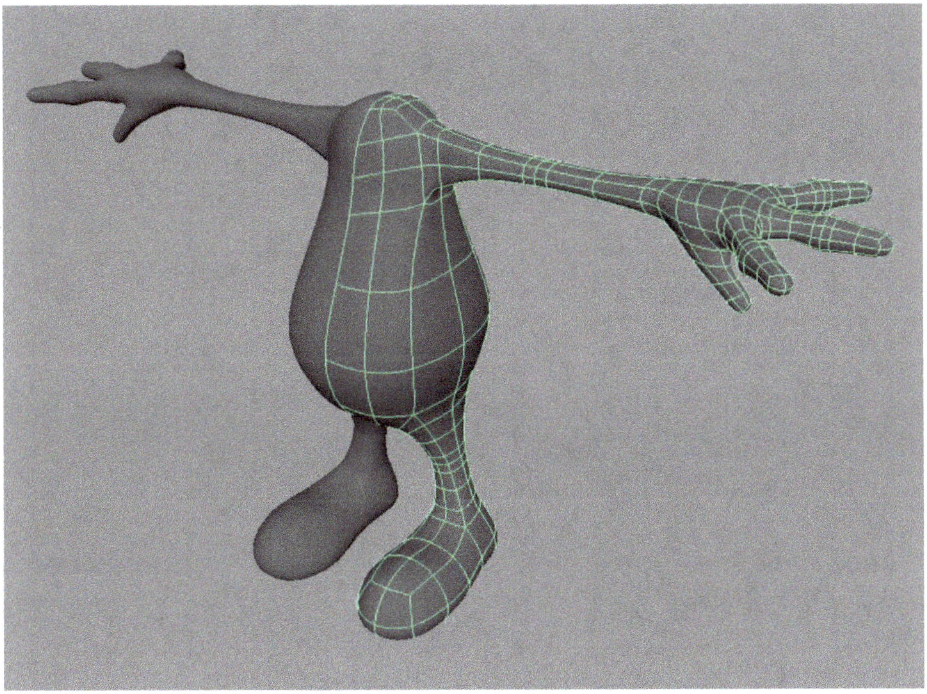

1.7 BLOCKING OUT THE HEAD

Now in the front view, select the top faces and extrude upward for the neck. Give it a division or two.

TIP

Before extruding you may wish to flatten the selected faces, making them coplanar, for more predictable results when they extrude. To do this, use the Scale tool and incrementally scale them in Y several times until they fall into a flat plane.

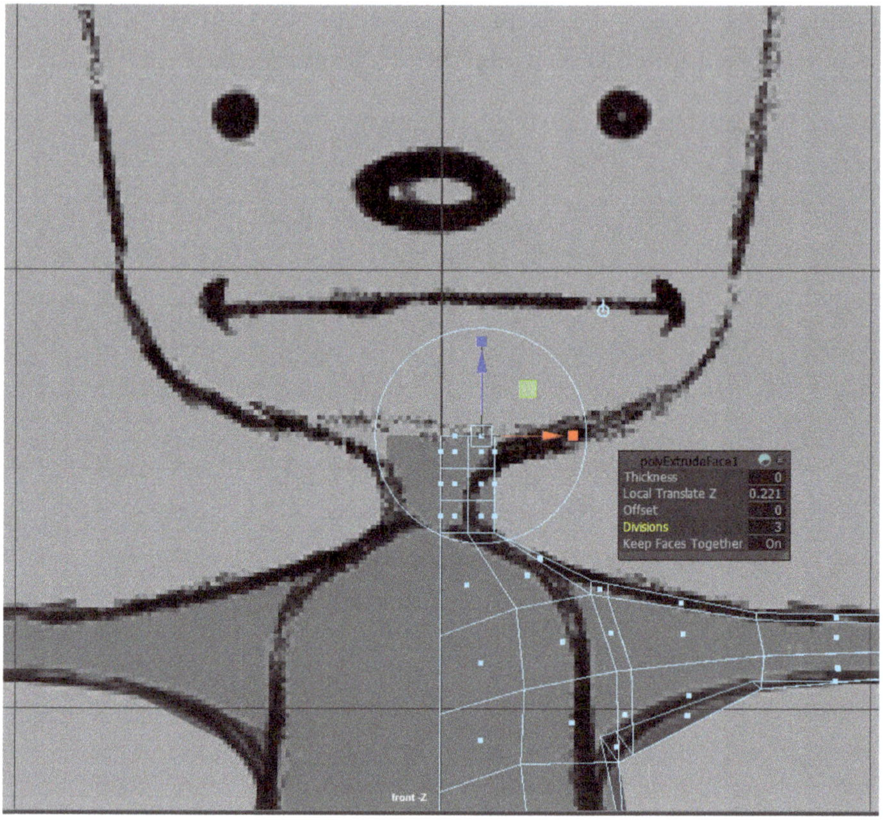

Hit **G** to repeat the extrude and bring it up to the chin area. Move the outer vertices to the width of the head in the front and side views.

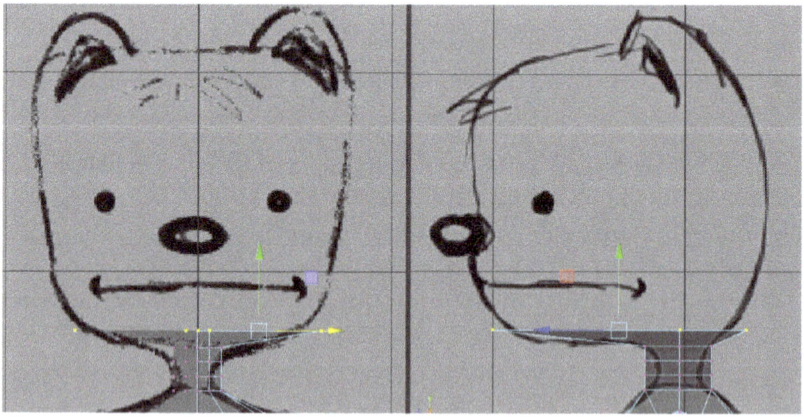

Extrude once more upward with 6 divisions.

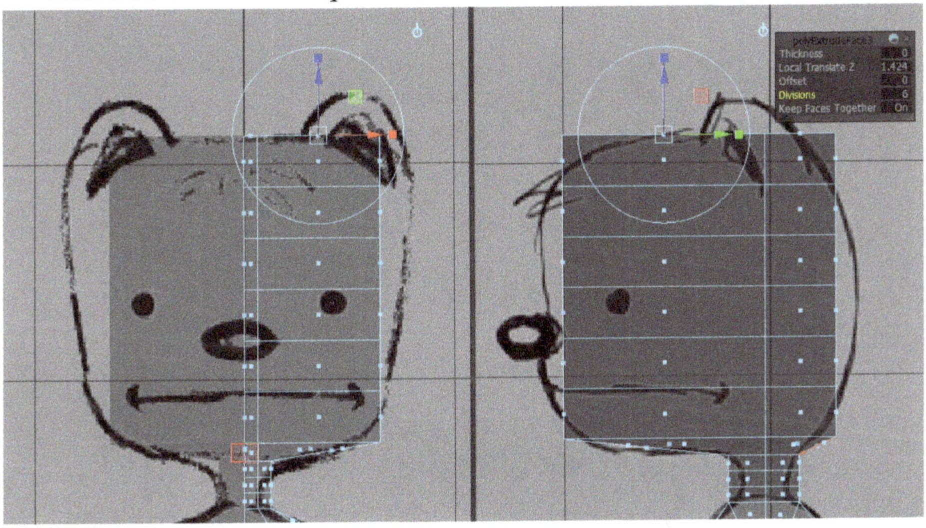

Don't forget to delete the hidden faces along the centerline!

Move the edge vertices to the silhouette of the character in the front and side views. Skootch the highest one inward to form the inside of the ear.

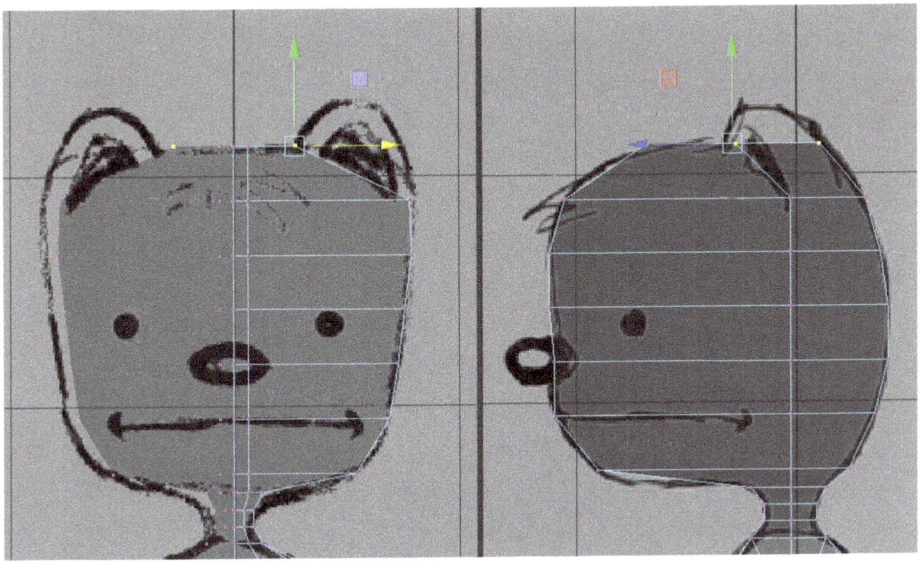

Select the corner edges and average vertices to round off the corners.

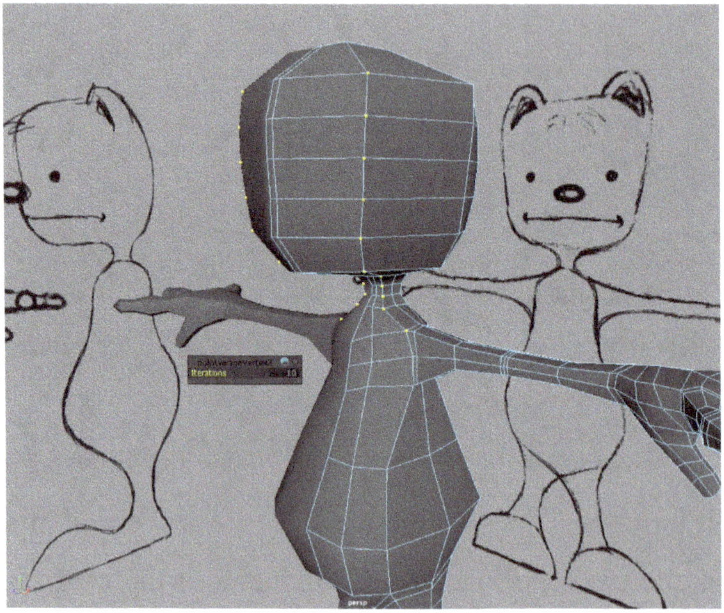

Select the backmost top face to extrude upwards for the ear. Divisions: 3.

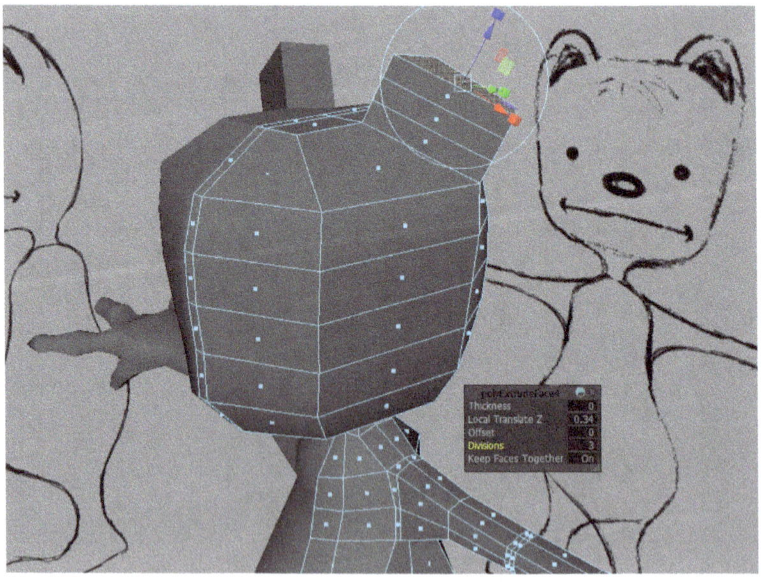

Select the top two edge loops and uniform scale them inward to approximate the tapered shape of the ear.

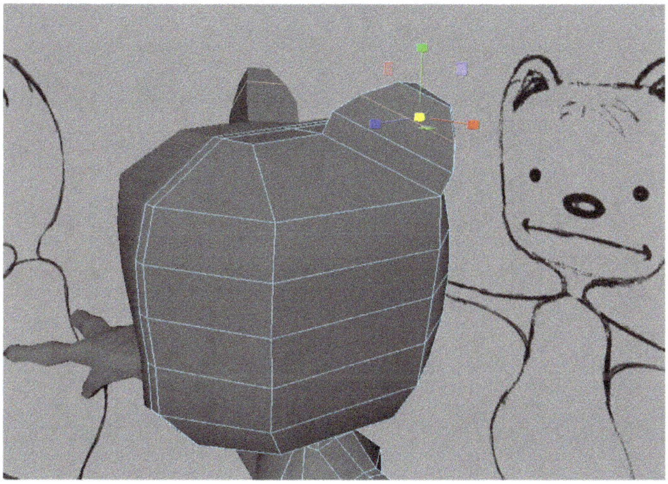

Select the front faces of the ear and extrude by scaling uniformly inward and then extrude again and push those faces back to make a cup-shaped ear.

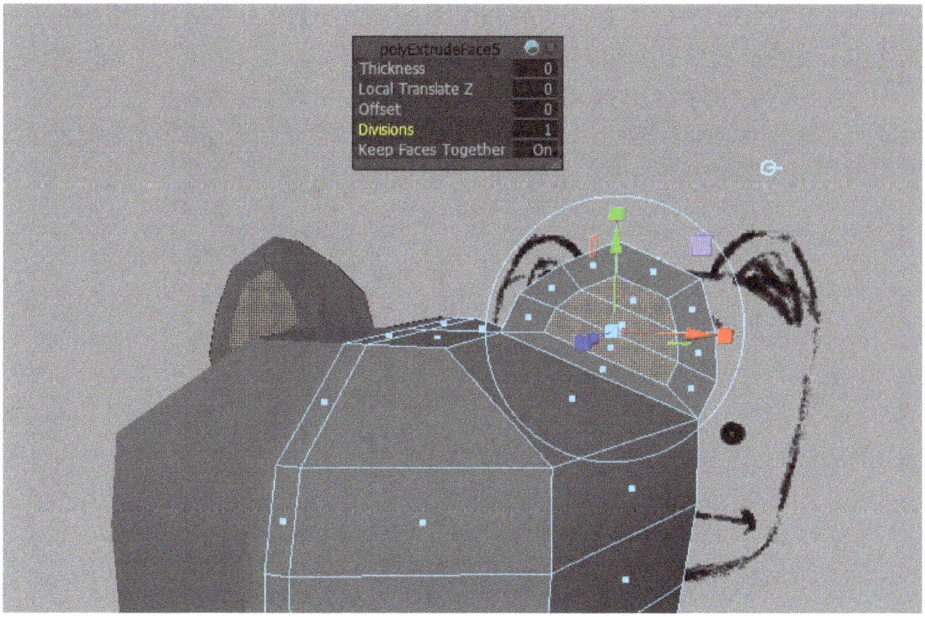

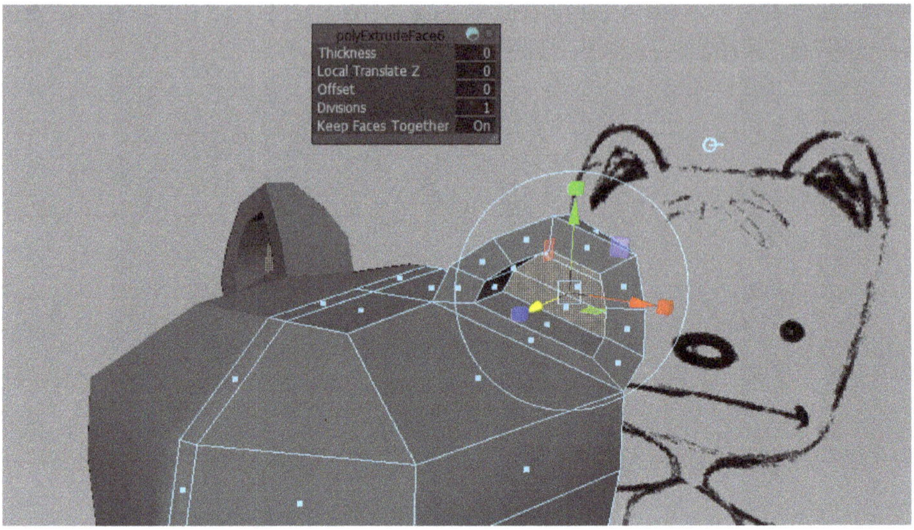

TIP

Use the local/world switch above the Extrude Manipulator tool to uniformly scale from all sides.

Use the Multi-Cut tool (**Mesh Tools > Multi-Cut**) to draw the following edge up the center of the face:

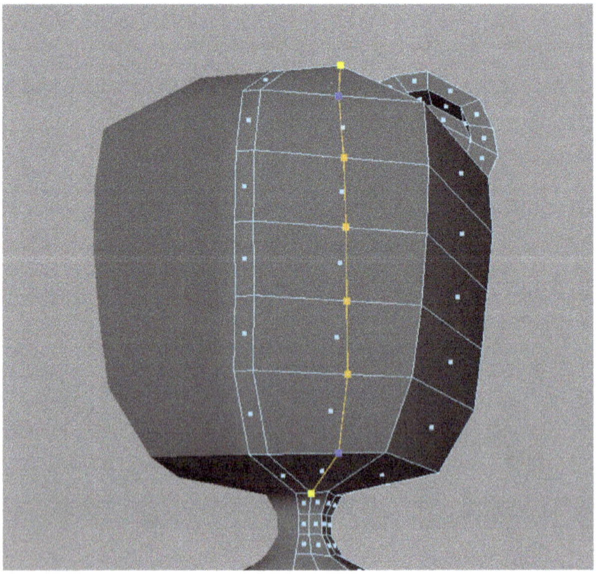

Then select the following faces and extrude scaling inward with three divisions to create a series of edgeloops around the eye sockets.

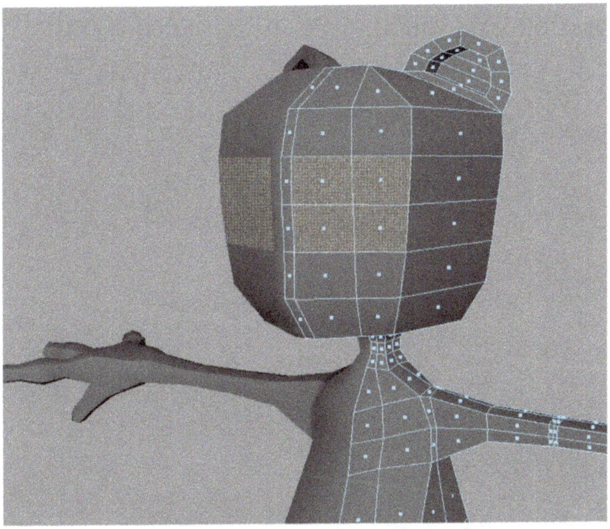

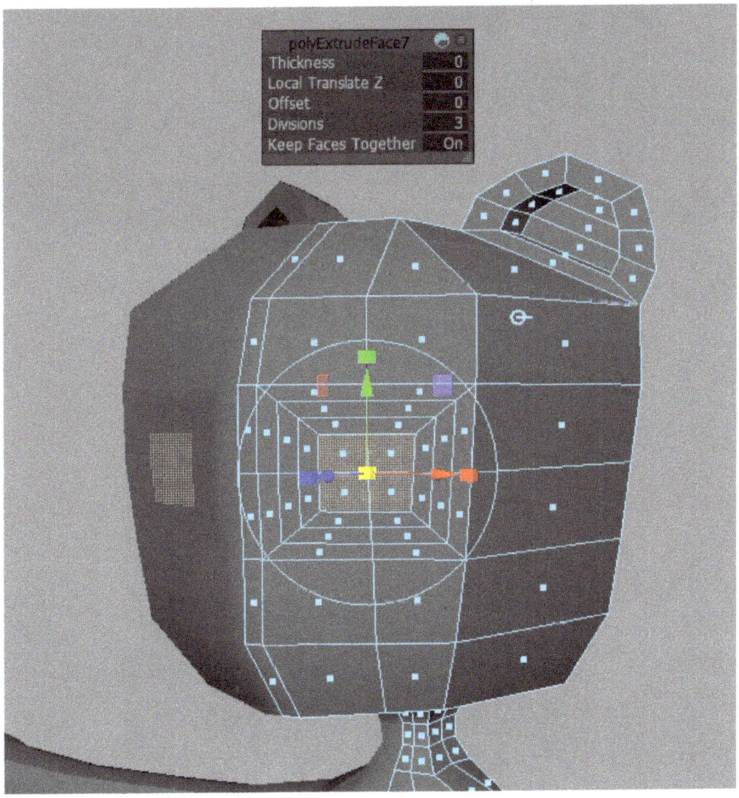

Do the same for the mouth, but remember to delete the inside faces and then snap the remaining verts to the centerline.

First, extrude using the manipulator's scale, rotate, and move abilities to keep the geometry more or less planar with the rest of the face.

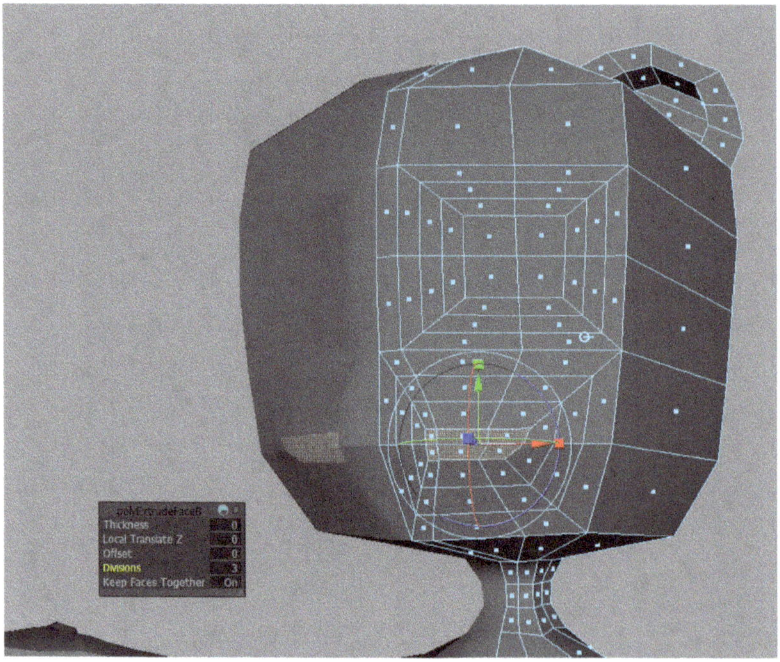

Next, delete the following extra faces:

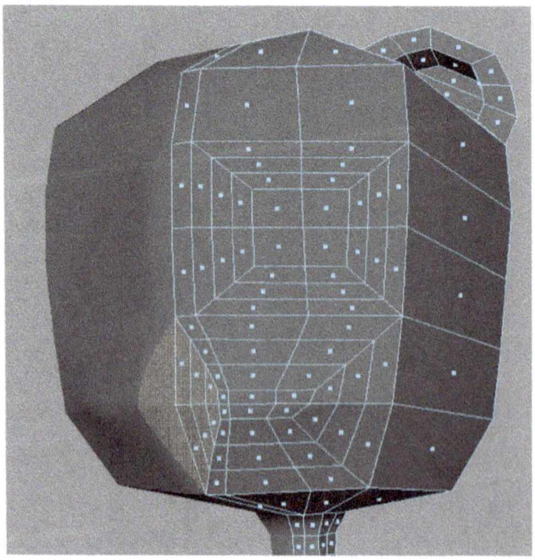

Select the remaining border vertices. To snap to the centerline, hold "x" to temporarily invoke "snap to grid" mode and move all selected vertices in X until they snap to the center. In the Move tool you will need to turn OFF **Move Snap Settings > Retain Component Spacing**.

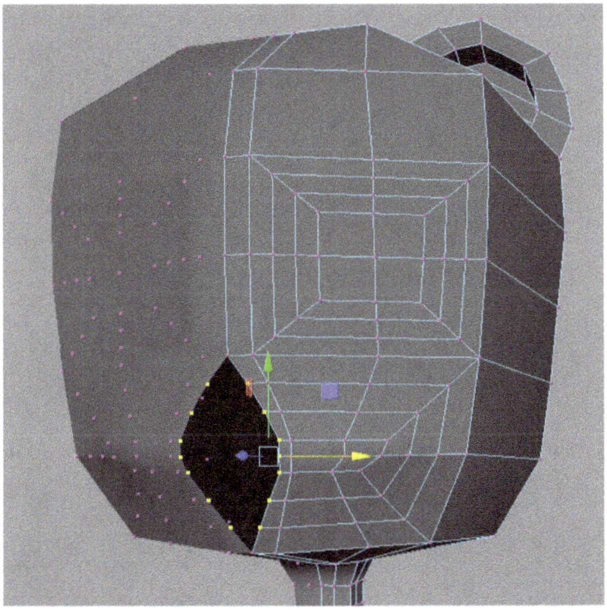

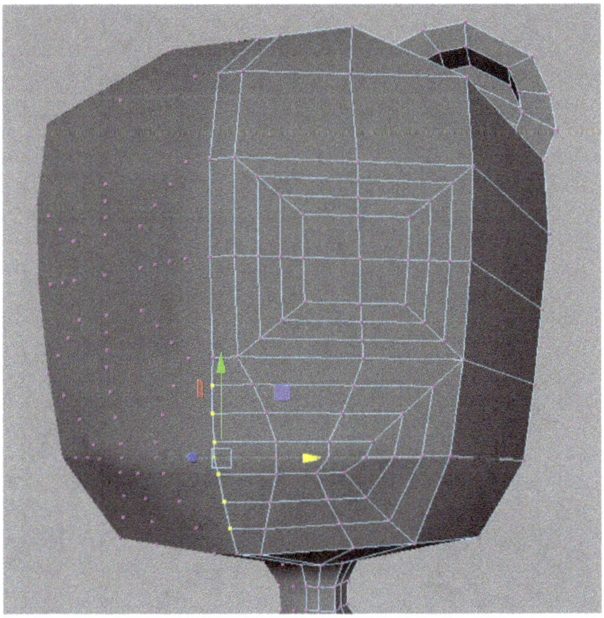

Adjust the remaining vertices to maintain a smooth profile.

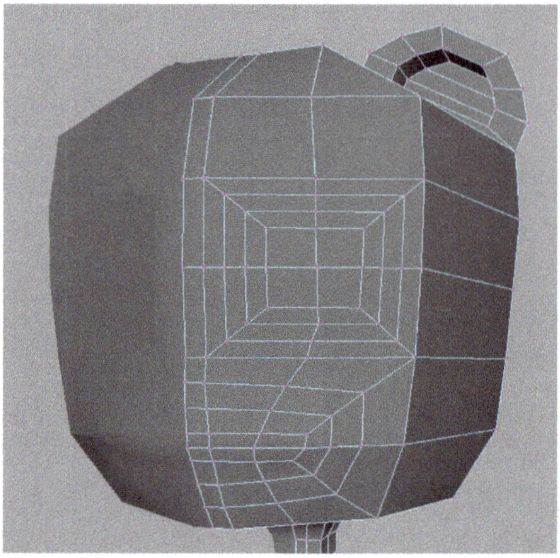

Extrude the innermost mouth faces back in Z to create the mouth sack.

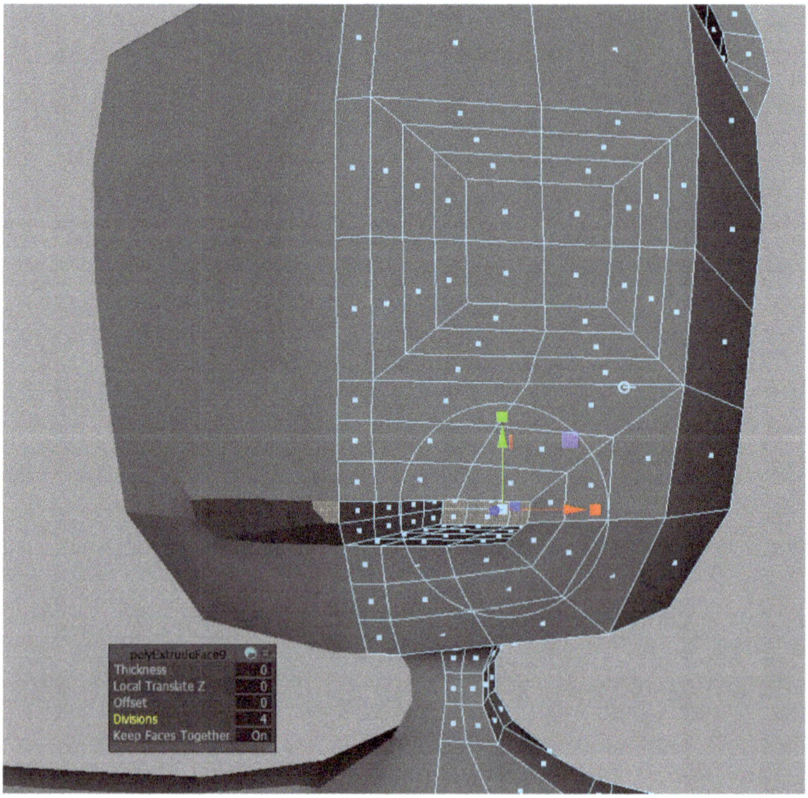

You can use Alt+H to isolate a selected half of the body to see what you're doing while working on the mouth sack. It can be relatively simple, but there needs to be geometry inside the mouth to cast shadows and hold the teeth and tongue that we'll build later.

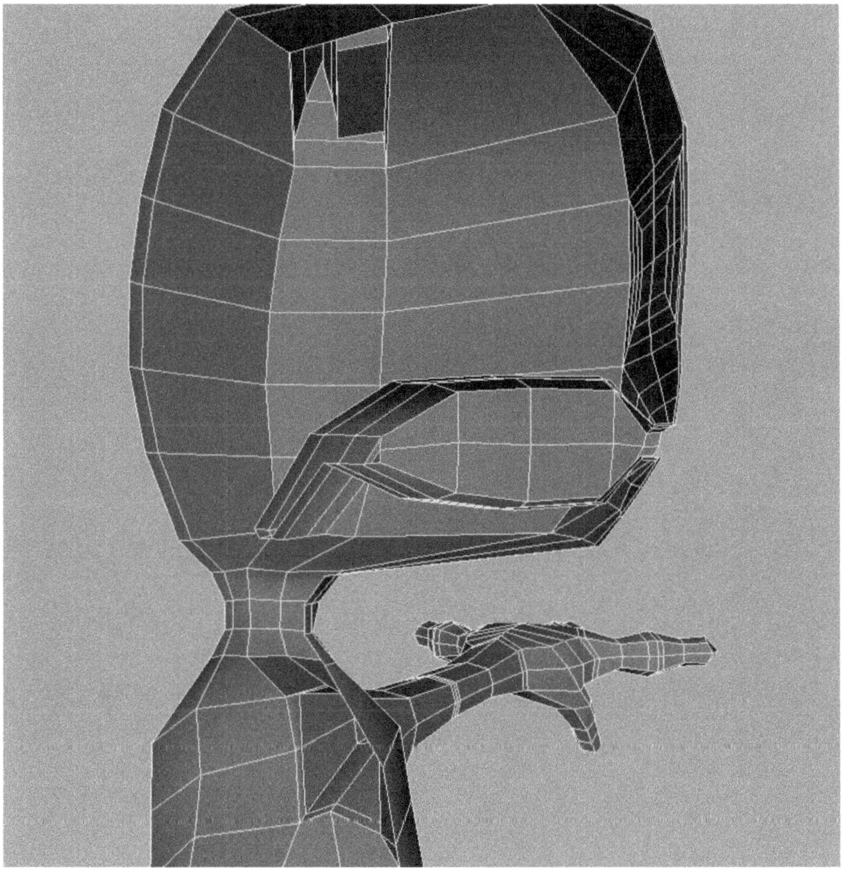

Don't forget to delete the faces extruded along the centerline. Leave one edge loop close to the lips to create an inside edge to the lips. Turn the back of the mouth sack down into an esophagus—no need to model beyond what will be seen when the mouth is open. Remember to keep all centerline vertices on the X grid center; at any time you can double-click a border edge to select the entire centerline and use our grid-snap trick to make sure they are all properly snapped. Use Ctrl+Shift+H to "show last hidden" to reveal the previously hidden half.

Now you have a *base mesh* for the Generikat. If you need to do any massaging to get all the verts into place, do it now. Use soft select, average vertices, and other tools to smooth and nudge every vertex into place based on the line drawings of your image plane.

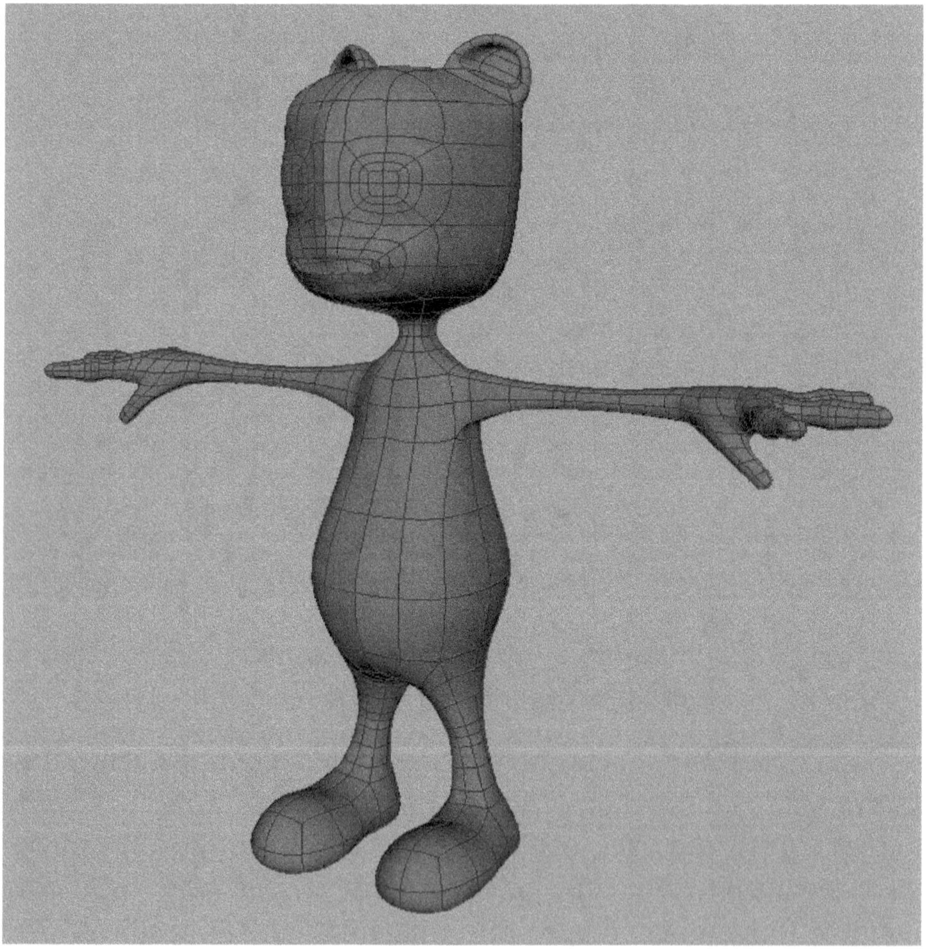

1.8 ADDING MESHES FOR THE EYES, TONGUE, AND TEETH

Create > Polygon Primitives > Sphere and switch to the Move tool. Hold **V** while you middle-mouse drag around the vertex defining the center of your eye ring (make sure the center of your Move tool manipulator is highlighted so it snaps in all three axes). The sphere will snap to the center of your eye geometry. Scale it down to a reasonable size and embed it into your head geometry by moving it back in Z.

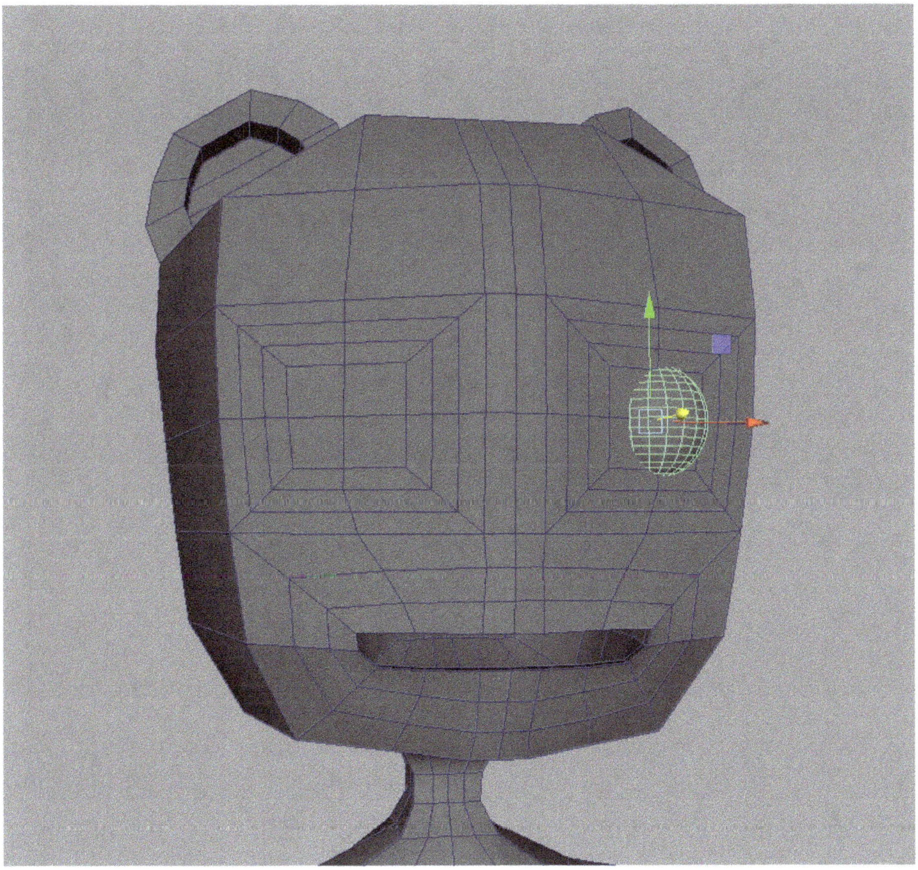

Use this sphere as a guide to model the edgeflow of your geometry around the eyeball.

TIP

You may want to template it while you work so you don't accidentally select it (right-click over it > **Actions** > **Template**).

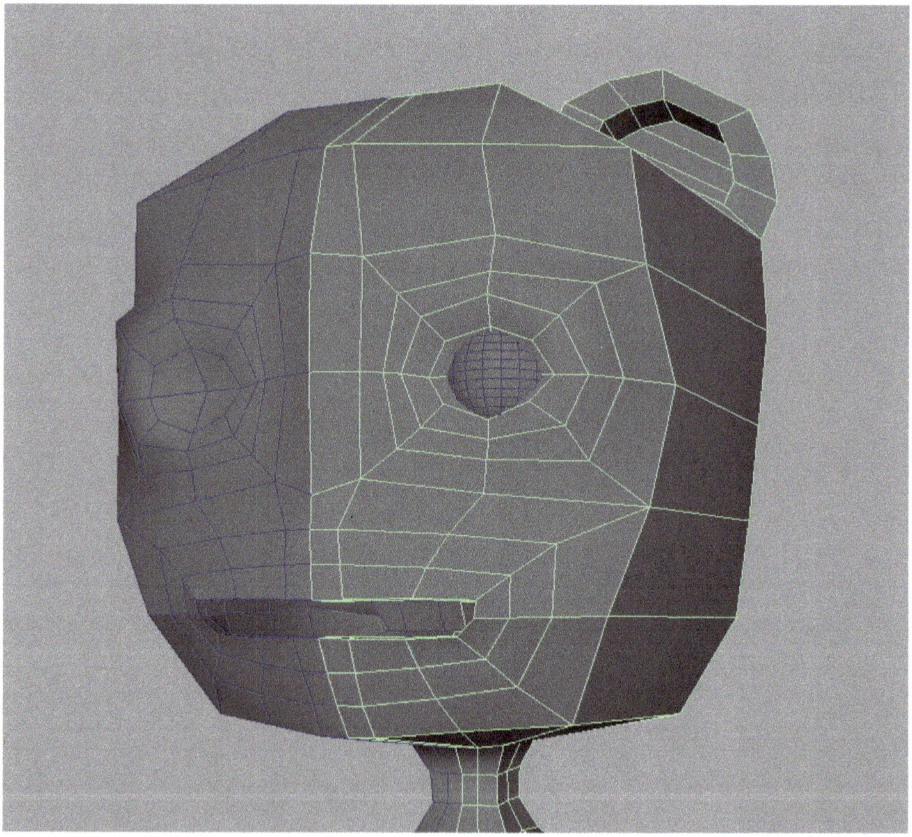

We'll make *clamshell* eyelids for this character. Toon-style characters can get away with eyelid geometry that is separated from the rest of the face for convenience in animating (allowing for blinks to be keyframed separately from eyebrow/facial animation).

In the front window, drag-select all of the faces above the horizontal center line of the eye.

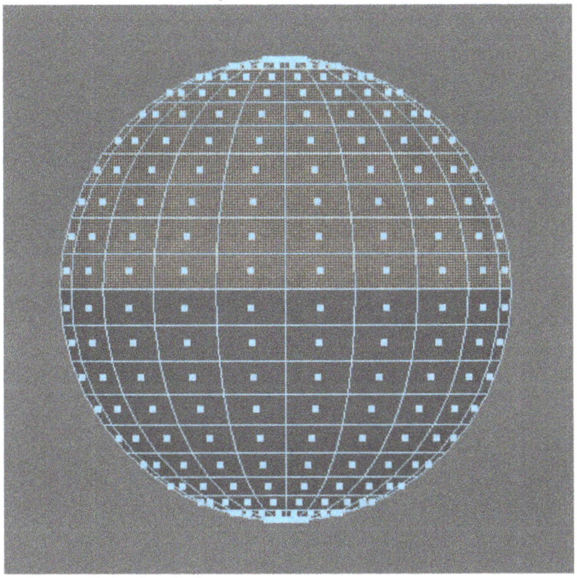

Edit Mesh > (Face group) > Duplicate. Uniformly scale the duplicated faces up to be slightly bigger than the sphere.

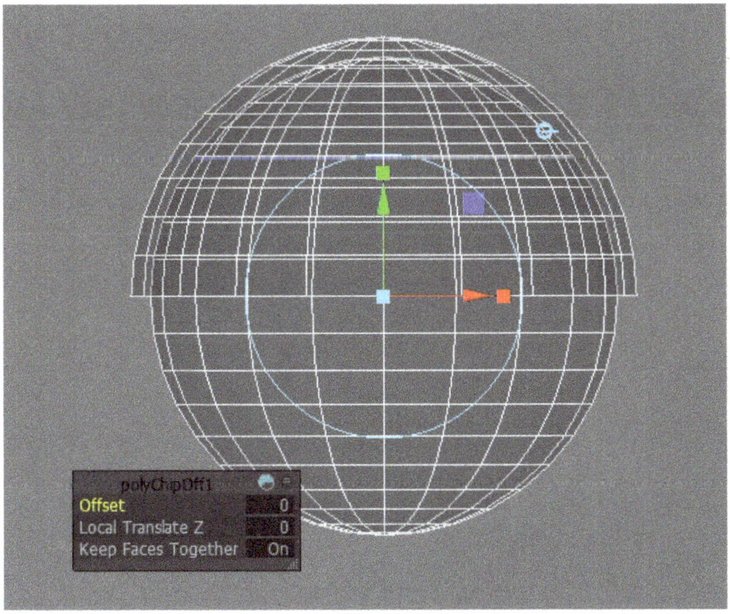

Select the eyelid geometry (at this point probably called *polySurface2* in the Outliner) and extrude it inward—or negative translate in local *Z*—so that it just interpenetrates the eyeball geometry. This gives the eyelid thickness.

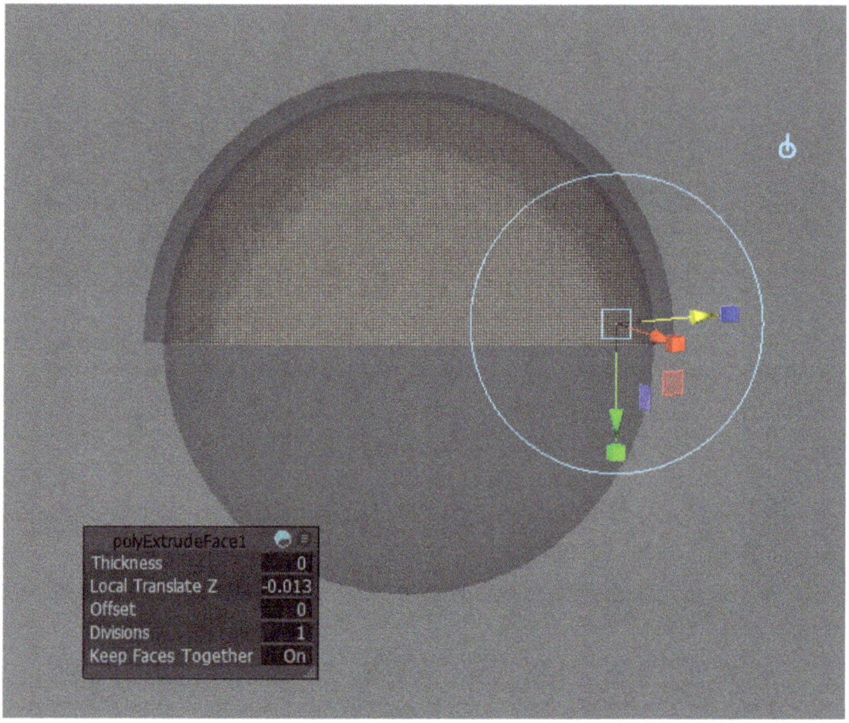

Edit > Duplicate Special, in the dialog **Edit > Reset Settings** and then change **Scale Y** to **-1** and **Apply**. This makes the bottom lid.

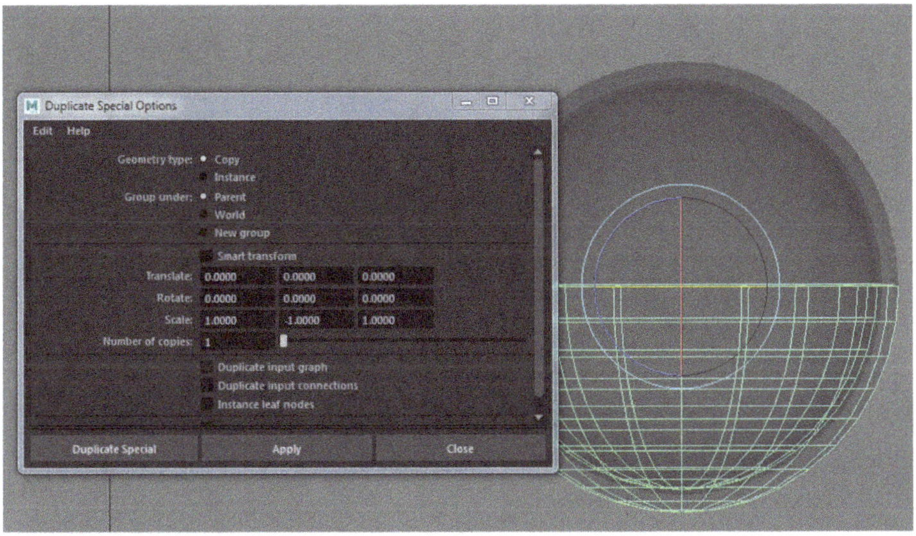

In the Outliner, select the top group node (probably called *pSphere1*) and **Edit > Duplicate**.

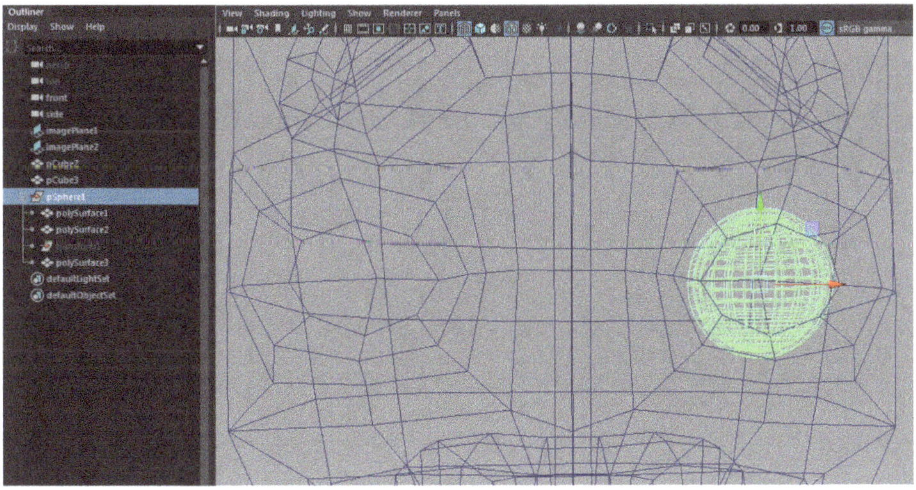

Select the duplicate Group (*pSphere2*) and put a negative sign in front of its **Translate X** value. That will pop it to the other side.

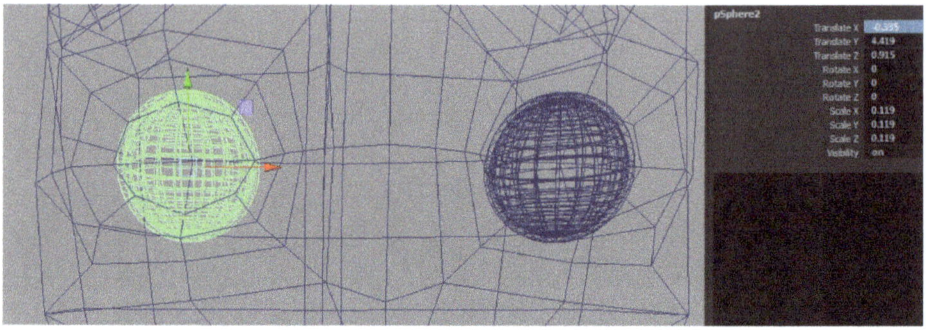

TIP

If the eyelids appear black in shaded mode with **Two Sided Lighting** turned off, **Mesh Display > Reverse**.

Select both groups and each node inside them and **Modify > Freeze Transformations** and **Edit > Delete by Type > History**.

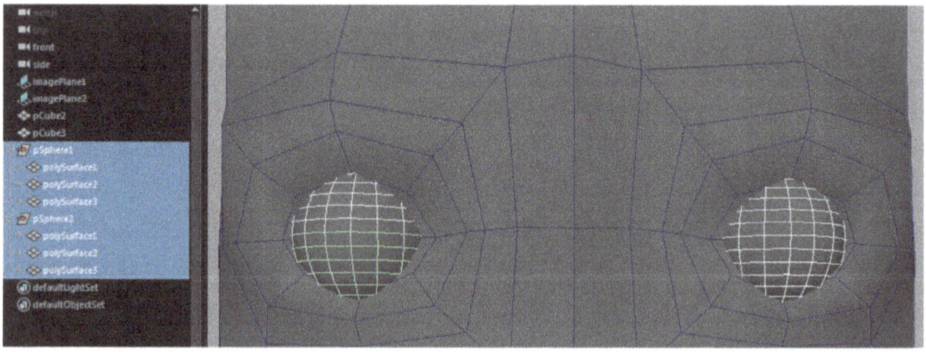

At this point you may wish to name these nodes for easy selection later. Double-click names in the Outliner to rename, or single-click the name in the Channel Box. Select each eyeball (inner sphere) and **Rotate X = 90** so that the pole is face forward.

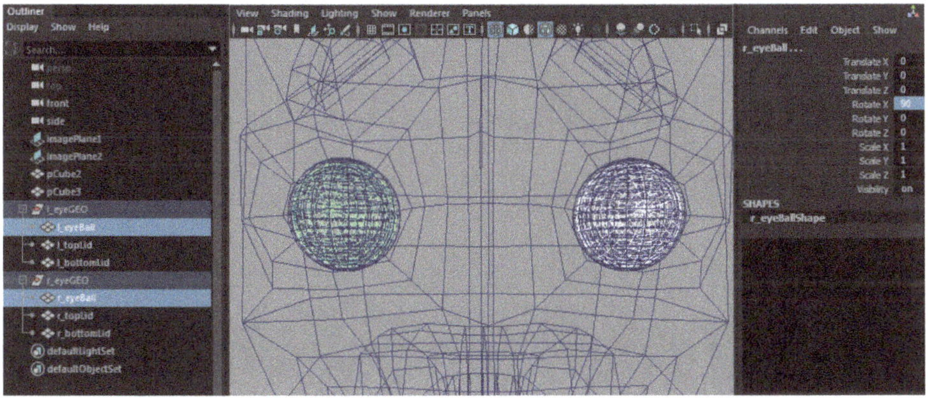

You can test the eyelids by selecting the Rotate tool and rotating them in X. They stay aligned with the eyeballs since their center pivot is the same as the eyeball. At any time return them to a closed state by returning **Rotate X** to **0**. Or, leave them open while you work so Generikat feels alive.

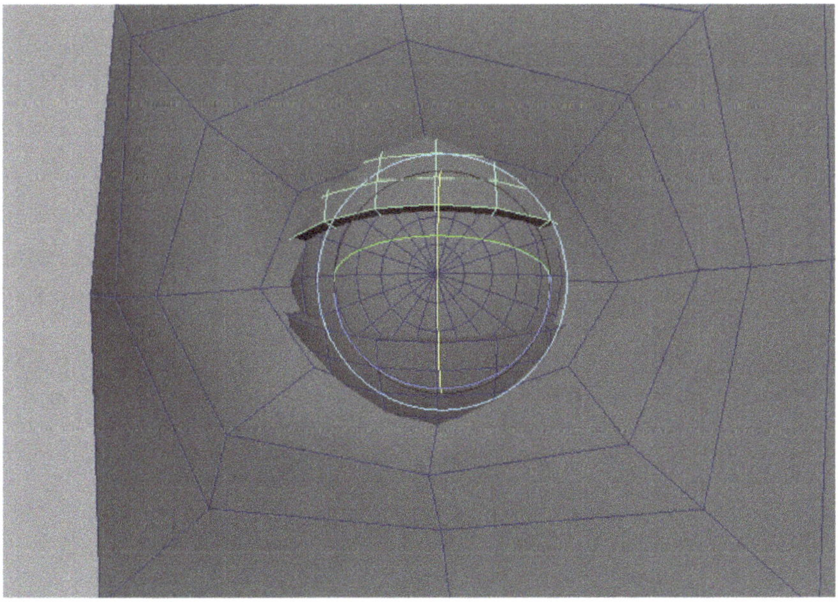

Create a nose with a sphere positioned at the centerline, halfway embedded in the facial geometry.

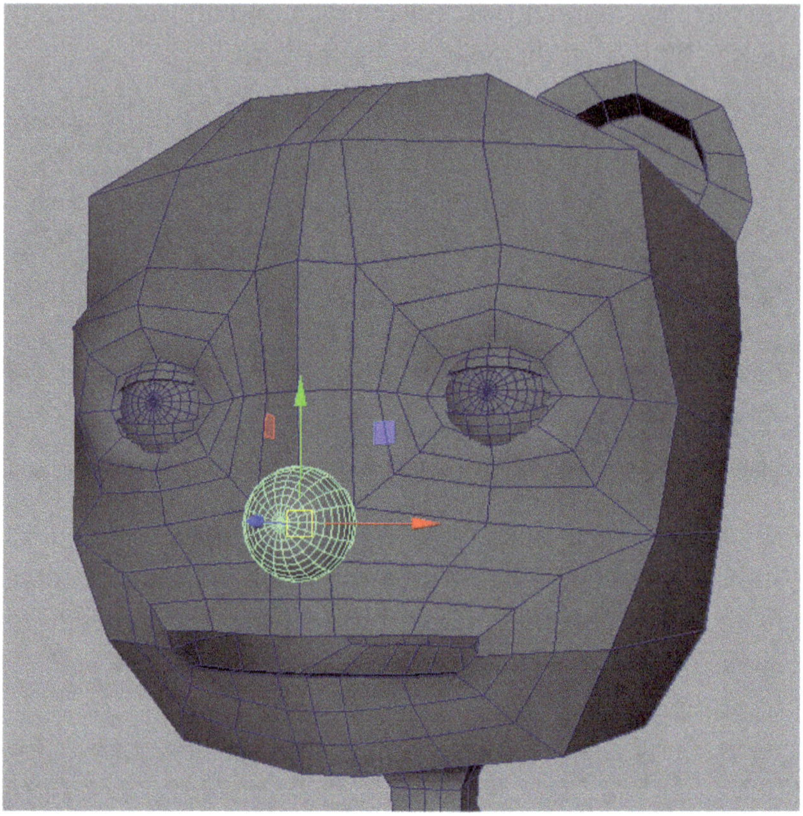

Now for the teeth and tongue. I keep a library of stock teeth and tongues to import on such an occasion. Every time you create teeth, tongues, or other reusable body parts, save them off as separate Maya files to import into future characters (**File > Export Selection…**). For now, we can quickly build simple ones from cylinders and planes.

You can fashion a tongue out of a 12x12 subdivision sphere by rotating the pole to face the front Z axis (rotate 90 in X) and then scaling into a flat, thin, horizontal tube. Bend the rear of the tongue to disappear down the mouth sack and move the top centerline down slightly for the tongue's center indentation.

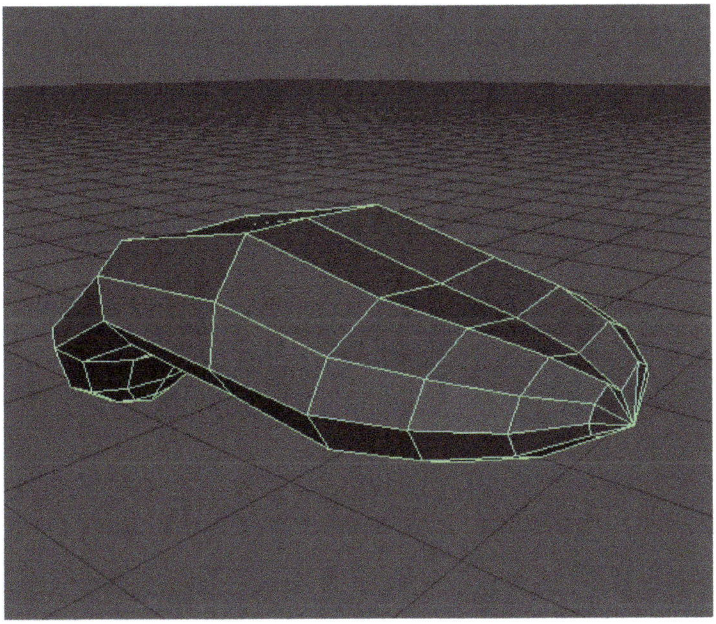

TIP

Soft Select (hit **B** with any transformation tool) is your friend when performing any of these kinds of modeling transformations. Just remember to hold B and horizontally drag the left mouse button to adjust the falloff radius.

Simple teeth can be easily fashioned from a pipe (**Create > Polygon Primitives > Pipe**). Delete the back half and fill the holes (**Mesh > Fill Hole**). Duplicate the teeth to create upper and lower and place them in the mouth sack just behind the lips.

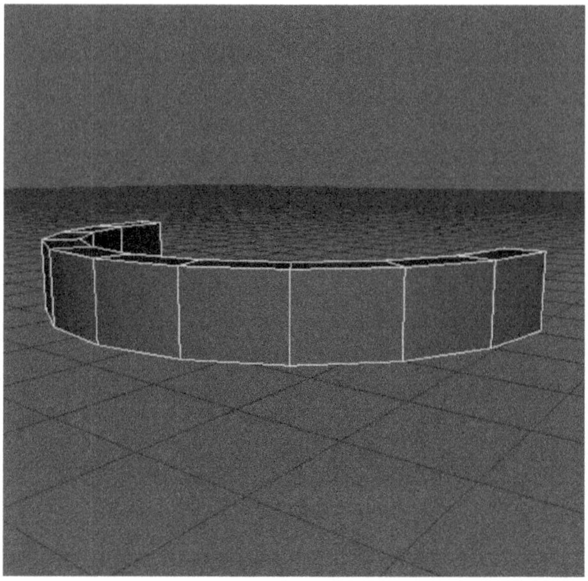

With the base mesh all blocked out, you are ready to up-res the model slightly. I added the following edgeloops to Generikat.

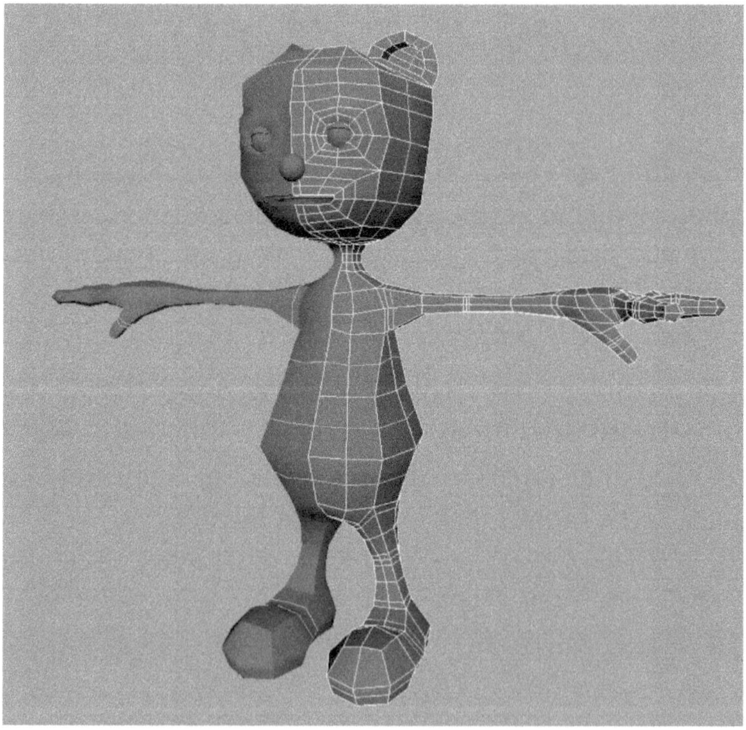

TIP

The Multi-Cut tool is handy to add edge loops, as you can hold Ctrl to add a loop and finish incomplete loops manually. To force edge loops to follow contours, you may have to flip edges of triangles (Ctrl+Alt+Left and Ctrl+Alt+Right) in places like the neck.

Now take some time to adjust vertices. The goal should be to define form while paying attention to topography: Quads should be as square and uniform as possible. See the following figure and notice some of the things I did: Knuckles and pads of fingers are suggested, mouth is closed with lips suggested, body is rounded off to create clean lines. A triangle edge at the top corner of the mouth was flipped for better edge flow, and care was taken to place the top-lip vertices across from the bottom. Knees, elbows, and shoulders would be suggested in a more realistic character. Feel free to deviate from my proportions to put some of your own style in it.

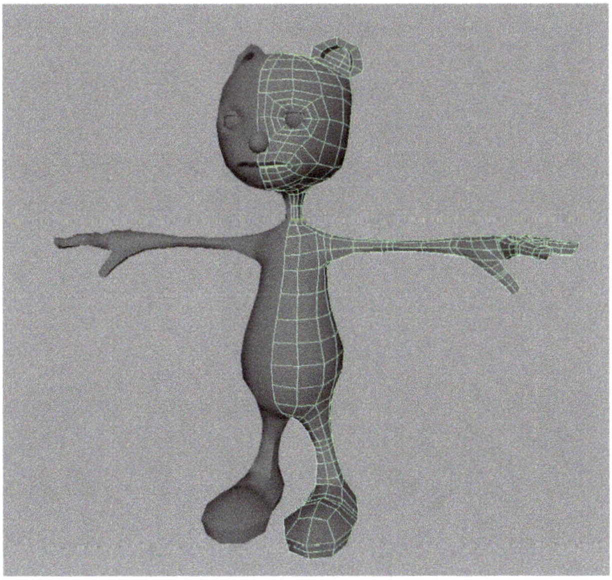

You are now ready to move on to UVing and then rigging!

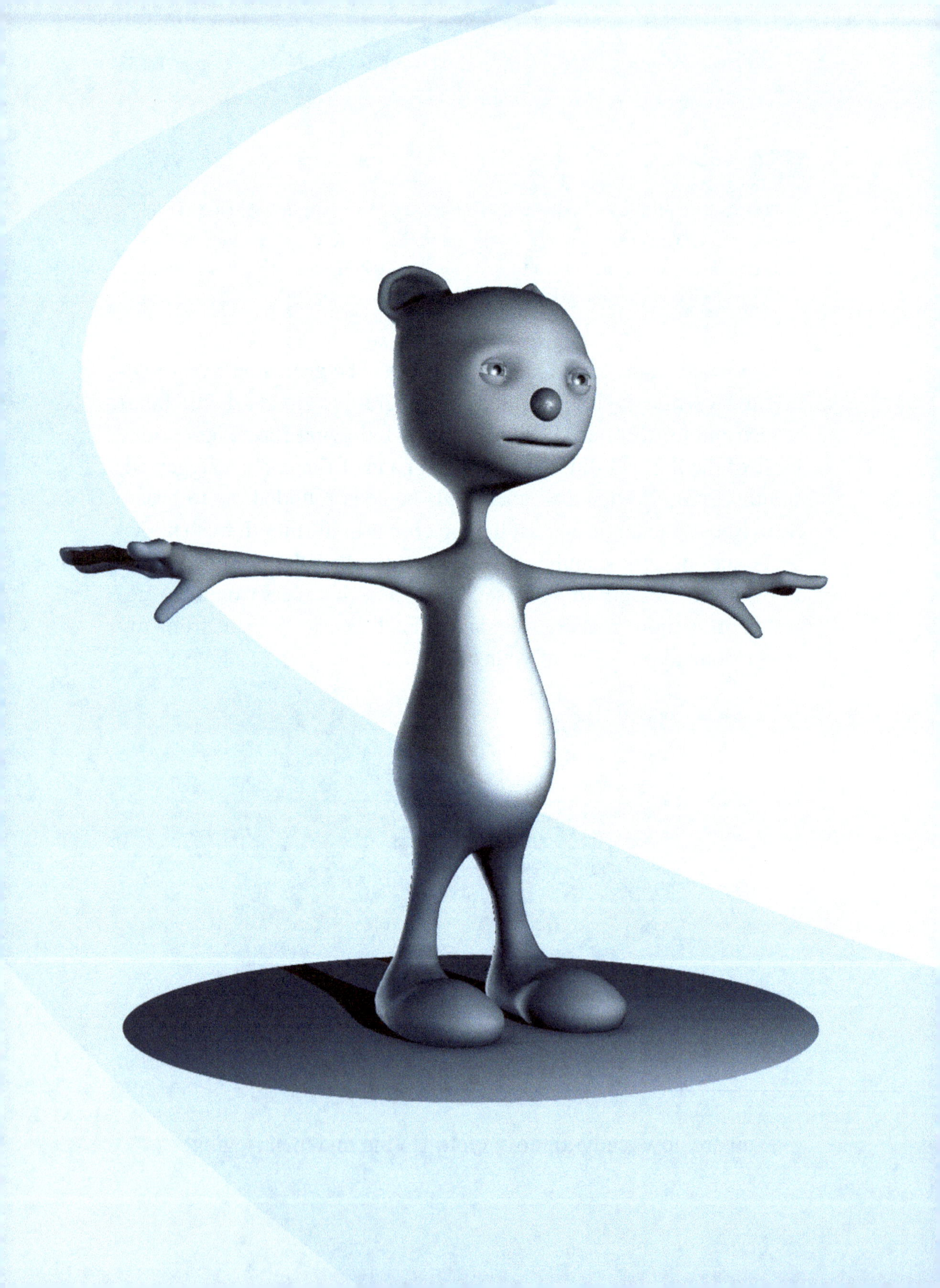

PART II

CHARACTER TEXTURING

2.1 CLEANING UP THE MESH

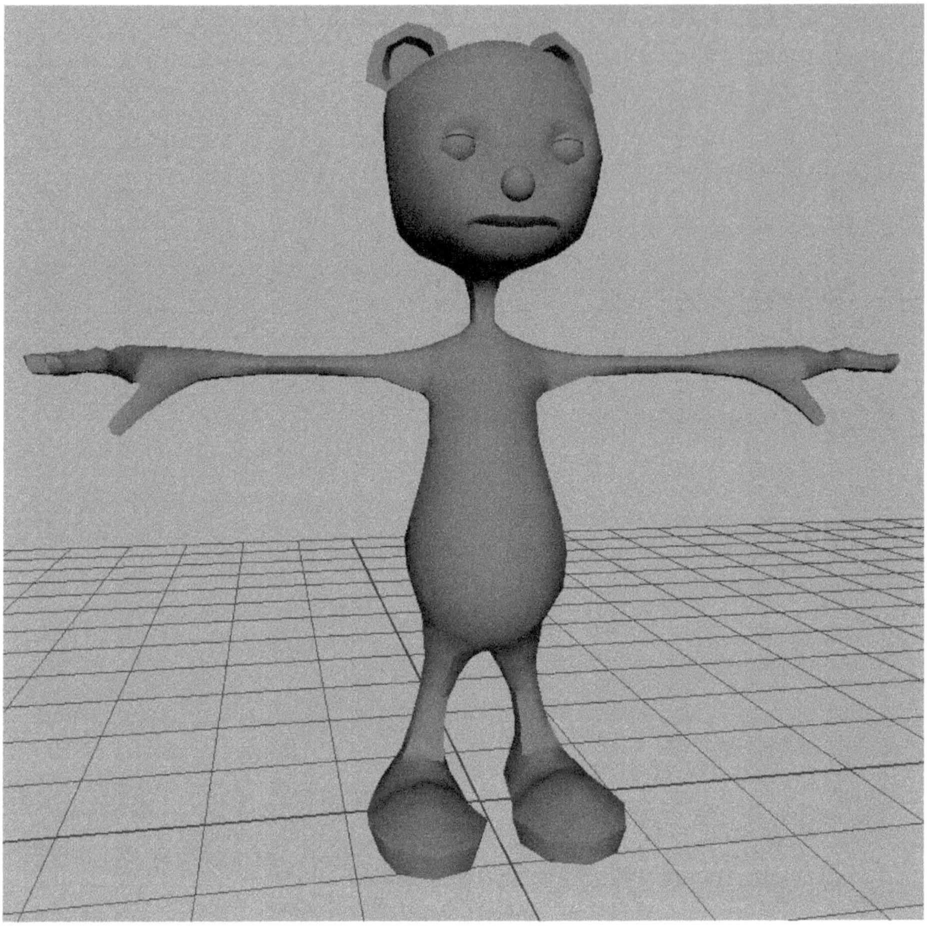

Welcome back. At the end of Part I, we had modeled a sort of teddy bear inappropriately named Generikat.

Before we start texturing and rigging there's some technical work to be done. First, delete the instanced half and, with the first half selected, **Mesh > Mirror** (with default options). This duplicates the geometry across the X axis and merges the border to make it a seamless mesh.

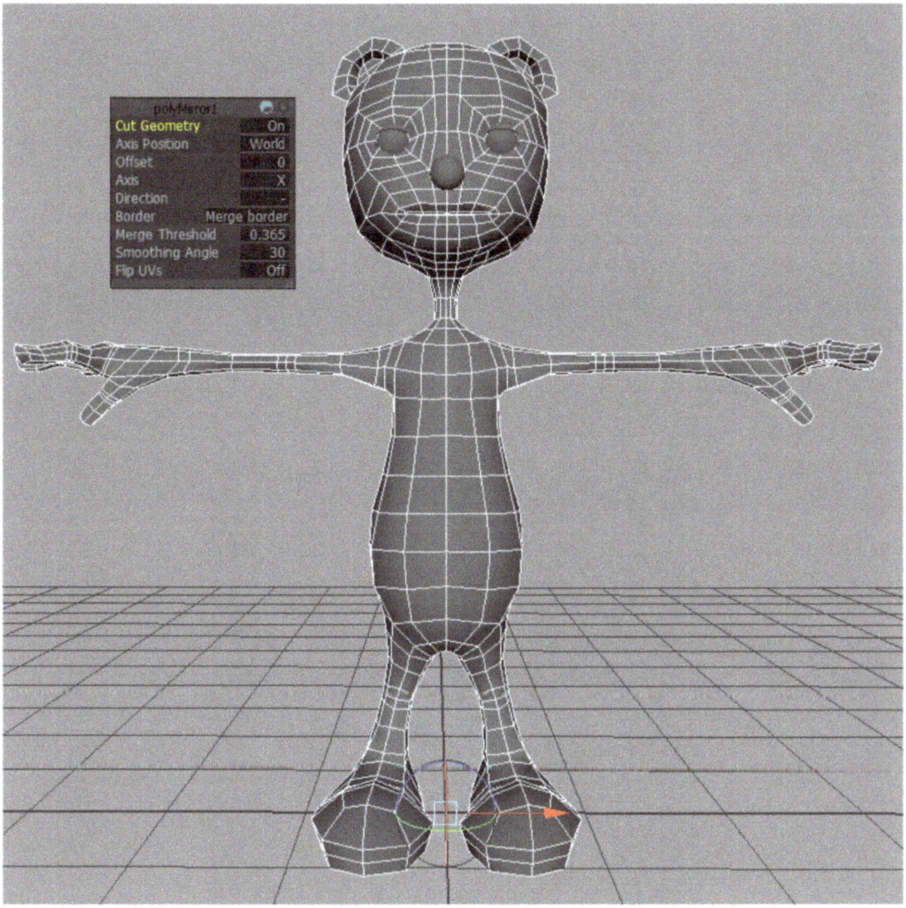

After creating the other half, you can more easily add features that straddle the centerline, like a little tail:

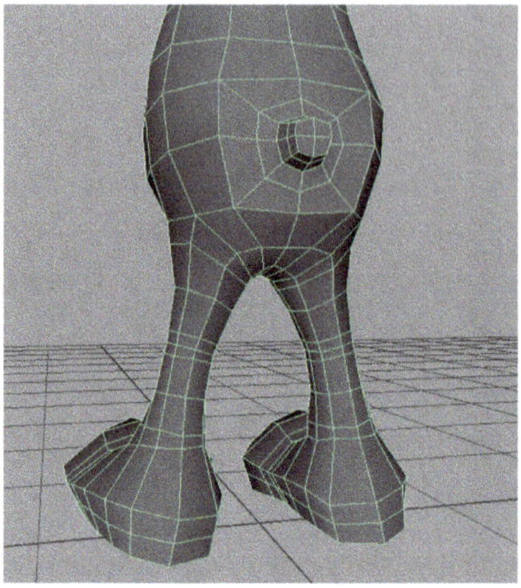

As a model check, run **Mesh > Cleanup…** Use the settings shown here:

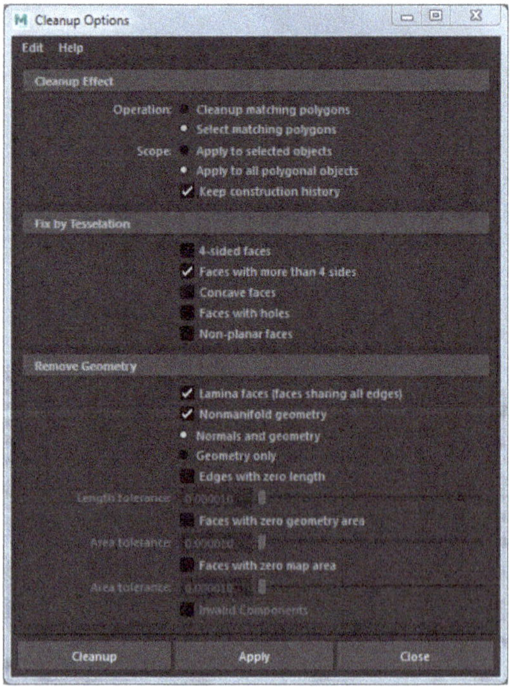

TIP

Note that **Operation** is set to **Select Matching Polygons**. Don't use the automated **Cleanup** operation; fix manually for much better results.

If anything comes up selected, you know you have some problem geometry to deal with.

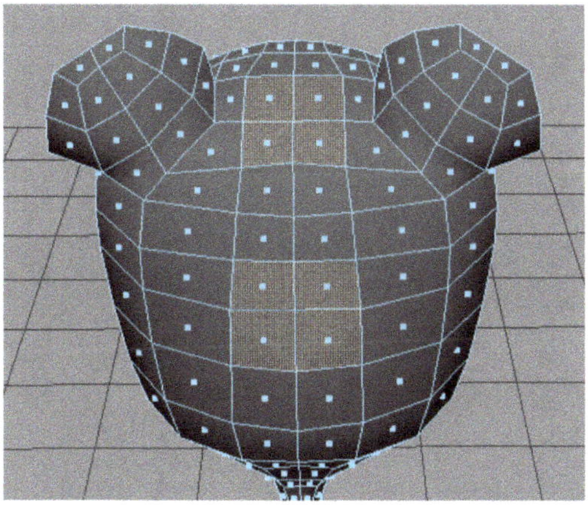

If faces are selected, look for non-quads caused by extra vertices. The Extrude tool and the Multi-Cut tool used carelessly are often culprits for these extra vertices. You can deal with them by **Edit Mesh > Delete Vertex** (Ctrl+Delete) or **Edit Mesh > Merge** or **Mesh Tools > Target Weld**. Clean up polygons manually with these tools and then run the Cleanup with the **Select Matching Polygons** operation again. Rinse and repeat until nothing is selected when you run the tool. After all this, **Edit > Delete by Type > History** on your geometry.

2.2 PROJECTING UVs

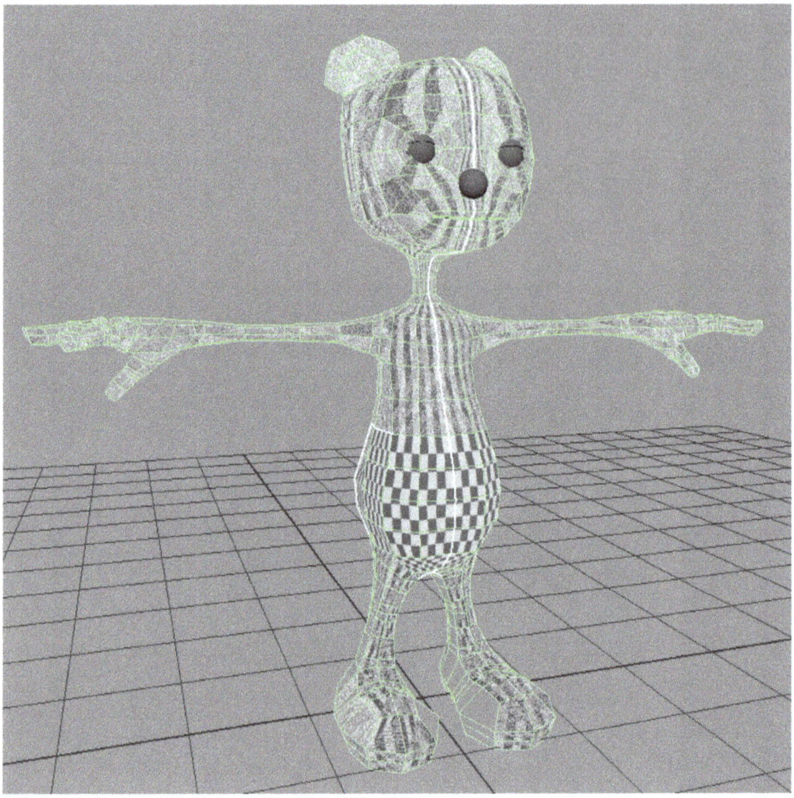

Before we can add textures to our critter, we need to lay out UVs, which are the texture coordinates for our model. Sometimes final textures are created after rigging, or even after animating, but UVs are generally laid out prior to rigging as the process can generate a lot of modeling history that we'll want cleaned up prior to rigging and skinning.

To see our UVs, open **UV > UV Editor** and select your model. In the UV editor, **Textures > Checker Map**. Chances are, the UVs look pretty bad. Since the box modeling process we've been using relies pretty heavily on extruding, and extrude operations do not auto-generate UVs, we've got whole areas of our model without UVs.

The first step of the UV process is to get some … any … UVs on there. **UV Editor > Create > Planar (options)**. In the options, **Fit projection to: Bounding box** and **Project from: Z axis**. You'll get one big UV shell that looks like an orthographic front view of our character.

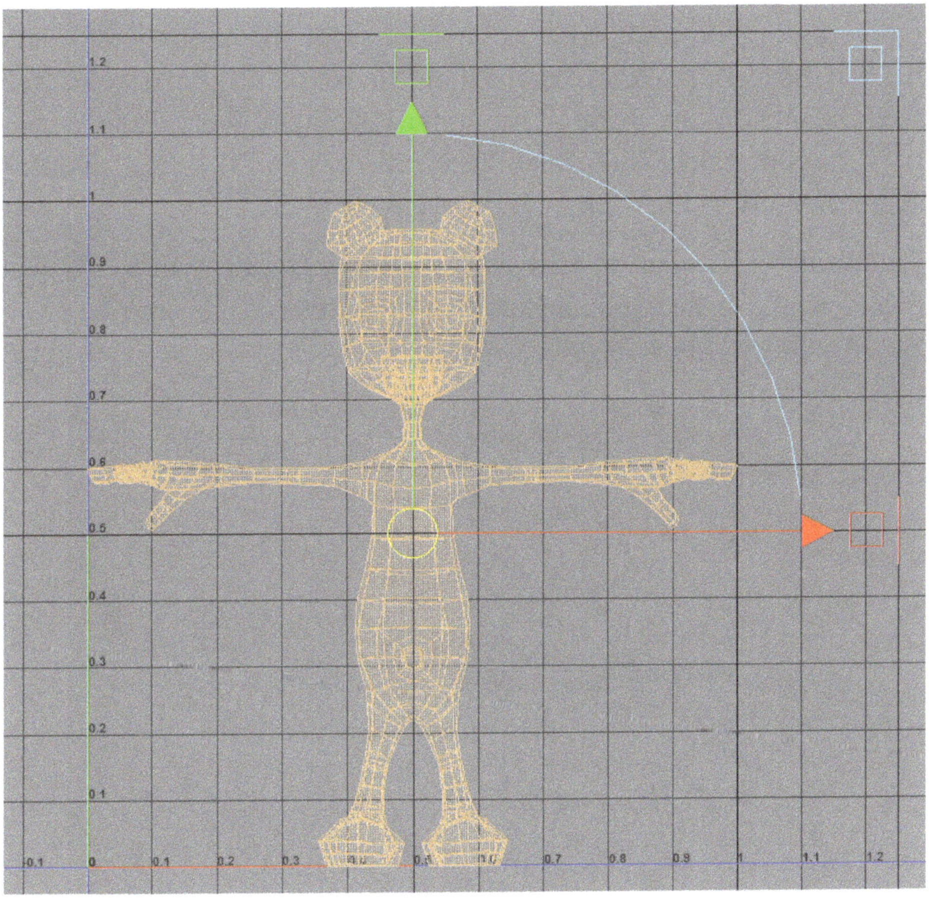

Now before we slice these UVs into islands (Maya calls them *shells*), make a selection set of all the UV border edges. By shift-selecting edges, select all of the edges that will make good seams for your character. (Double-click to select an entire edge loop.) Think ahead to how it will unwrap and put in enough edges to minimize distortion, all the while trying to keep the number of UV shells low and placing seams where they are least likely to be noticed (occluded areas, areas of low texture detail, or "natural" borders between different textures). In general, you'll want to cut off protuberances like limbs and ears by selecting an entire edge loop around their base, where they connect to the main trunk. Each limb will need a seam running lengthwise to unfold from. Place these seams on the inside of the limb where they won't draw attention. In your select tool, turn on **Symmetry** in **Object X** to auto-select the other side's edges.

When you have all of your edges selected, **Create > Sets > Quick Select Set...** so that you can return to this selection if need be (with **Select > Quick Select Sets**). The following images show where I put mine.

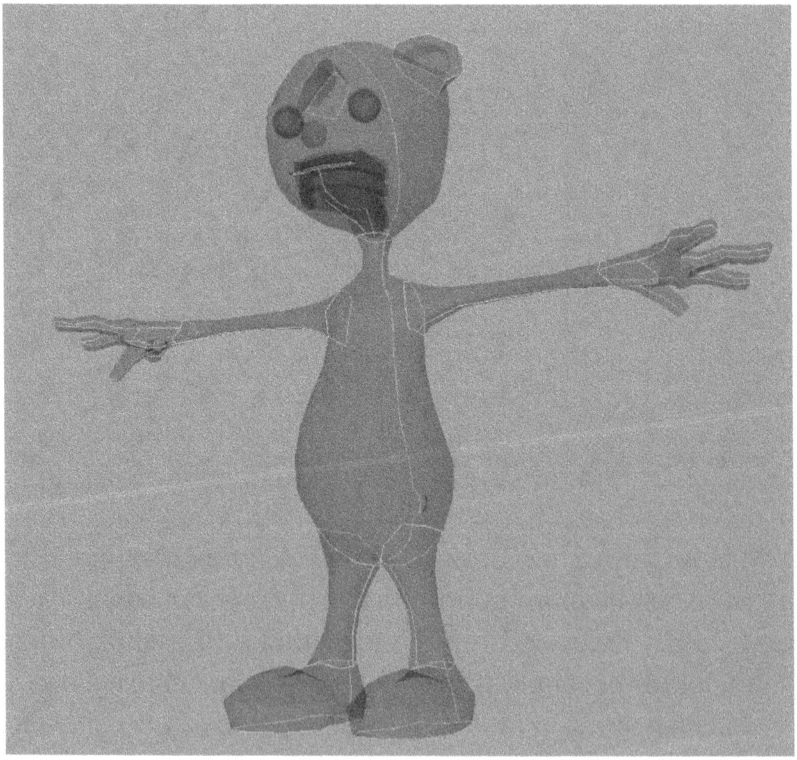

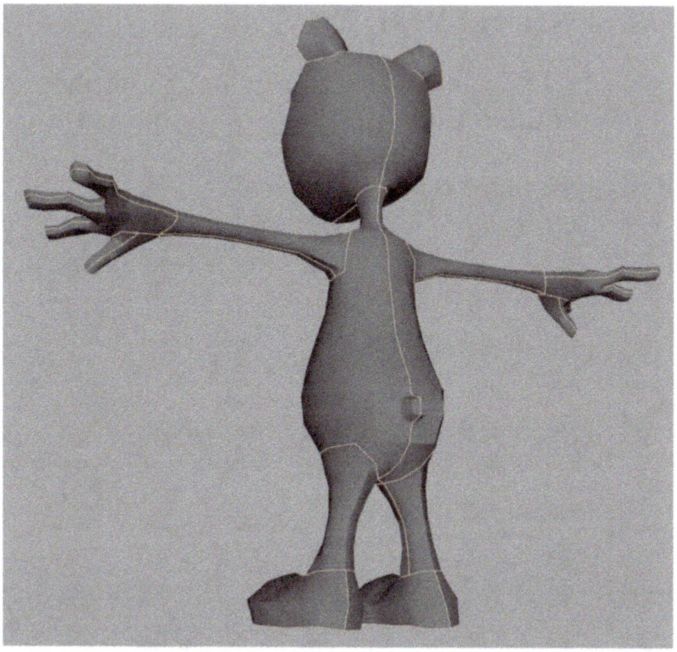

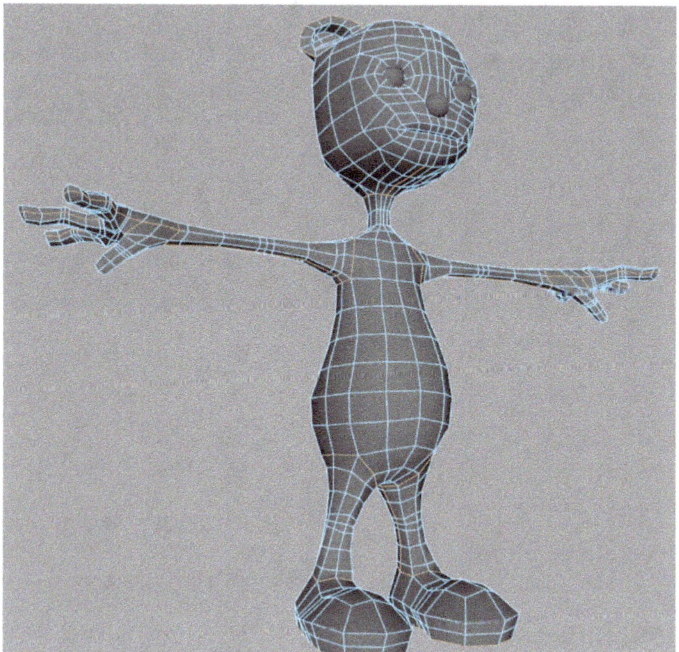

2.3 CUT AND LAYOUT UVs

Then go to the UV editor and **Cut/Sew > Cut** or Shift+X. You can use the **Tools > Move UV Shell Tool** to move the shells apart.

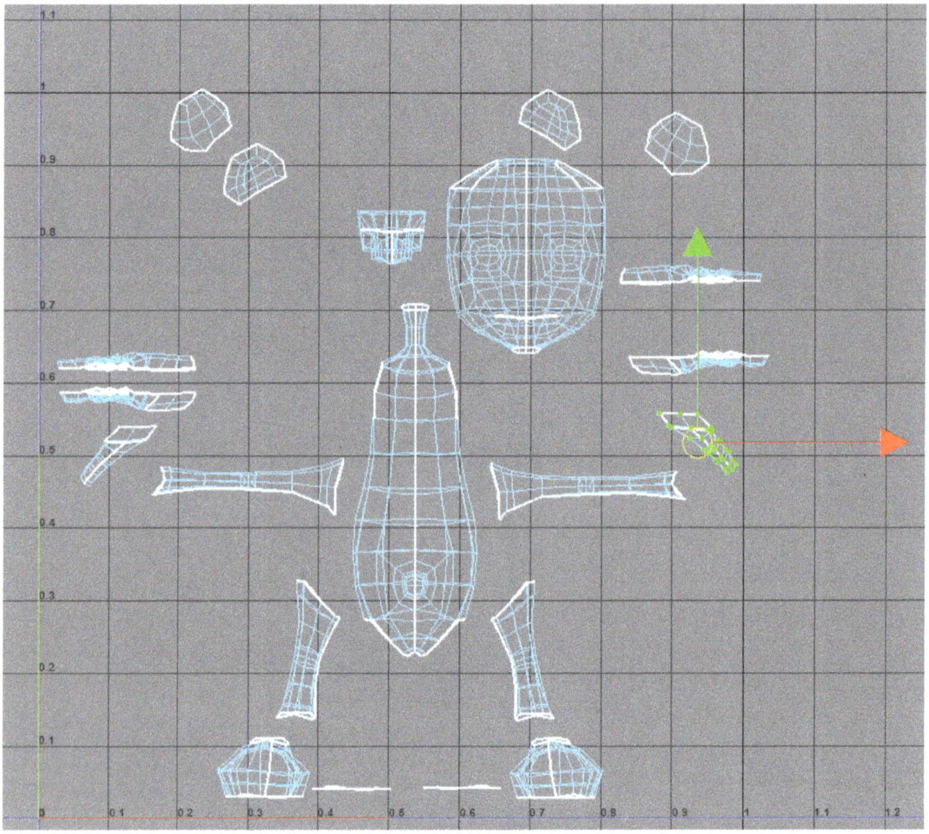

With every piece selected, do a **Modify > Unfold** and then a **Modify > Layout** to get results similar to below.

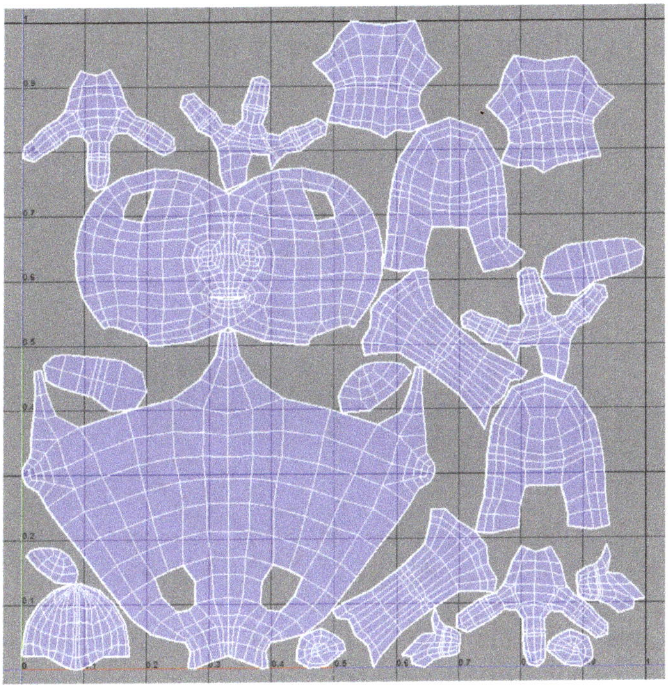

TIP

If **Unfold** is not giving desired results, try switching to **Method: Legacy** in the options, and weight toward a **Global solver**.

If some shells have crumpled or overlapping faces, scrub them clean with the **Tools > Optimize** brush.

At the minimum you should have every piece non-overlapping, as we do. If you'd like to take it further you can rotate, move and scale each shell to maximize texel space. Allow a few pixels of spacing between shells and tile borders (this is called **Shell Padding** and **Tile Padding** in the **Layout UV** options). The tools in the **UV Toolkit** (which opens when the UV editor opens) are your friends in this endeavor.

UVs should generally be of uniform scale. To ensure uniformity, select a shell with a good texel density (based on the size of the checker texture on the model in the viewport) and use **UV Toolkit > Transform > Texel Density > Get**. With this value, you can now select all your shells and hit **Set** to set them all to a uniform scale.

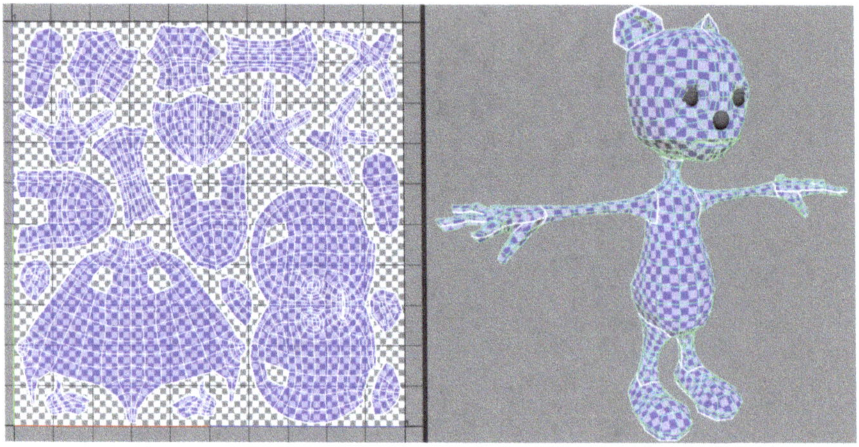

UV layout is an art form unto itself. Programs such as Headus UVLayout have a suite of tools that allow for much more detail-oriented UV layout. The strategy depends on considerations such as organic vs. hard body modeling or if you are texturing for games or film. If you rely on Photoshop for texture creation, you'll probably want to keep shells oriented upright and edgeloops straightened or snapped to grid lines. If texture space is a consideration (as in if you are exporting to a game engine) you'll want to be much more sensitive to maximizing use of texel space. For our purposes, since we'll be 3D painting a simple diffuse map, non-overlapping and relatively uniform UVs are our primary considerations.

If you check the UVs for your eyes, nose, and tongue, you'll notice they are fairly usable as-is. That's because they were created from polygon primitives with no extruding involved, so they still have their default UVs. Even the teeth are usable for a fairly plain shader. But if you want to give the teeth better UVs with minimal fuss, use **UV > Automatic** mapping, which gives non-overlapping fairly uniform UVs on simple objects.

TIP

If you have Maya Bonus Tools installed (available on the Autodesk website as a free download), try **Bonus Tools > UV Editing > Auto Unwrap UVs Tool...** It lets you choose border edges and then the tool automatically does everything else, to hit-or-miss results.

2.4 MATERIALS AND TEXTURES

Now we'll add materials to our geometry. If you haven't yet, name all your geometry and delete history. Naming becomes very important when we get into textures because we want a naming scheme that allows for models and textures to have similar but not identical names (Maya does not allow two scene objects to have the exact same name). I use a simple naming convention of the convention *objectNameTYPE*, with an abbreviated object-type token in all caps. For example, *noseGEO* for nose geometry, *l_eyeGRP* for left eye group, and *noseTEX* for the texture map that goes with *noseGEO*. (For multiple maps, you could use *noseDIFF* for diffuse, *noseSPEC* for specular, etc.) Whatever you do, don't use spaces in object names in Maya (Maya was originally developed for a UNIX operating system, which doesn't allow spaces in filenames). Instead, use camelCase or under_scores.

When you close your UV editor, the temporary checker shader goes away. We'll add new materials to all geometry—never start texturing on the default grey lambert, or you will affect all new objects in the scene. RMB over your *bodyGEO* and in the popup menu choose **Assign Favorite Material > Lambert**. This will attach a lambert material to this model called *lambert2*. Give it an identifiable color and rename it *bodyMAT*.

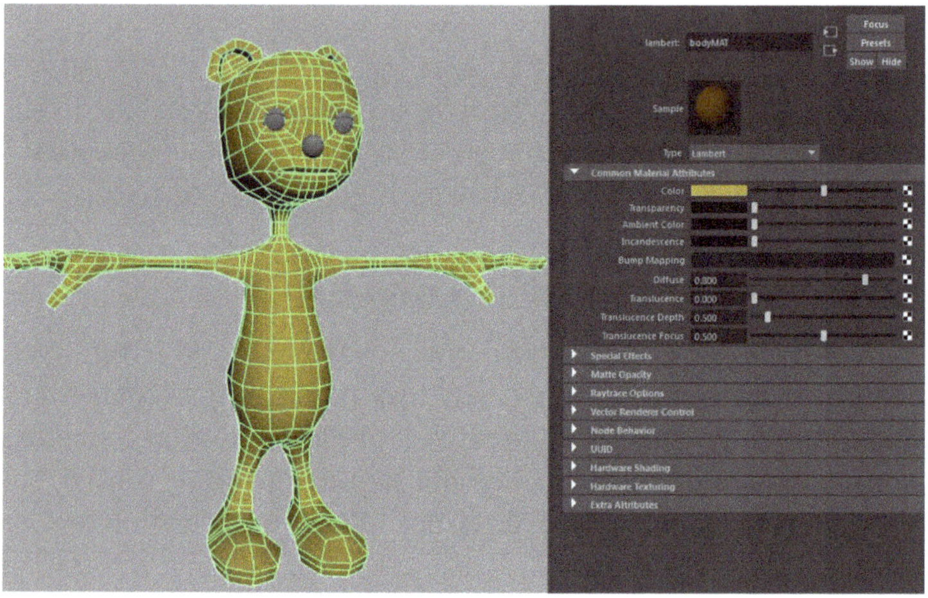

Select the eyelids and give them a lambert named *eyelidMAT*. For their color, you can click the swatch to open a color chooser. Use the eyedropper to pick the same color as the body (hint: use **Lighting > Use Flat Lighting** from the viewport menu to choose the unshaded color).

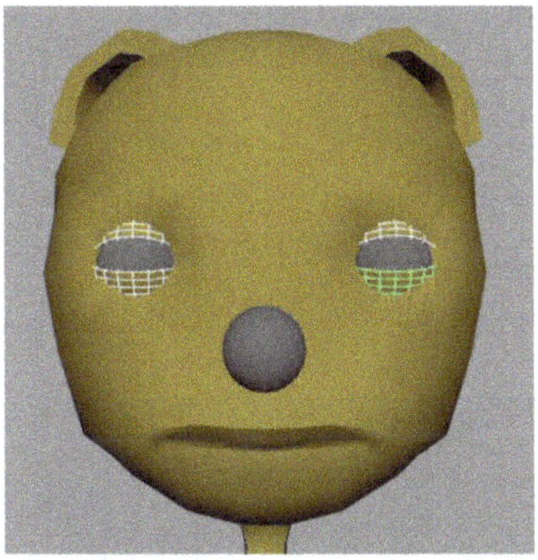

For the nose, add a Blinn. This material type has a specular component so the nose can look semi-glossy. I gave mine a dark grey color. The **Specular Color** attribute can be mapped to a texture by clicking the checkerboard icon next to its name in the Attribute Editor. In the **Create Render Node** window that pops up, choose **Fractal** or **Noise**, a procedural texture to roughen up the nose. You can adjust the attributes of the fractal to your liking, or even add it to the bump mapping slot as well. Note that Hardware Texturing **(6)** must be on to see textures in the viewport.

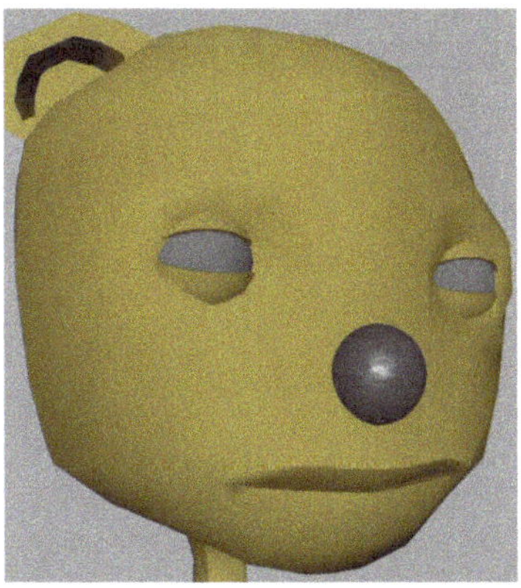

With the eye geometry selected, right-click over it and **Assign New Material…** Choose a Phong and map a ***ramp*** texture to its color swatch. In the Attribute Editor for *ramp1*, under **Ramp Attributes**, change **Type** to ***V Ramp*** and select and move the color pots above the ramp to adjust the placement and falloff of the ramp colors. Clicking anywhere in the ramp gradient will create a new color entry. Adjust the colors such that they appear as a pupil, iris, and sclera. Note that **Selected Color** can also be mapped; in the example below I mapped a ***fractal*** texture to the sclera to create veins and I mapped a ***noise*** texture to the pupil with the **Implode** attribute cranked up to create the streaks of the pupil.

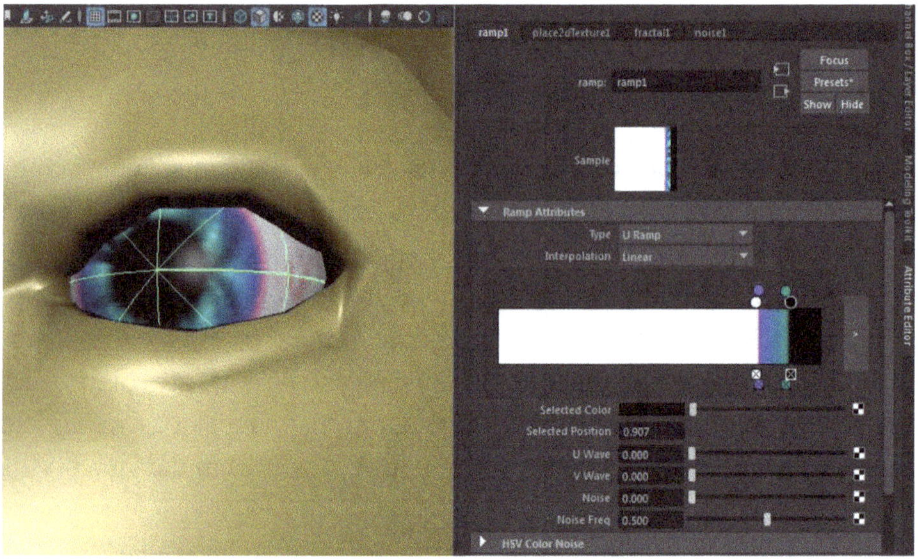

RMB over the other eye and choose **Assign Existing Material... >** ***eyeMAT*** to assign the same shader to the other eye. Isolate select (Ctrl+1) the teeth and tongue to create Blinn materials for them.

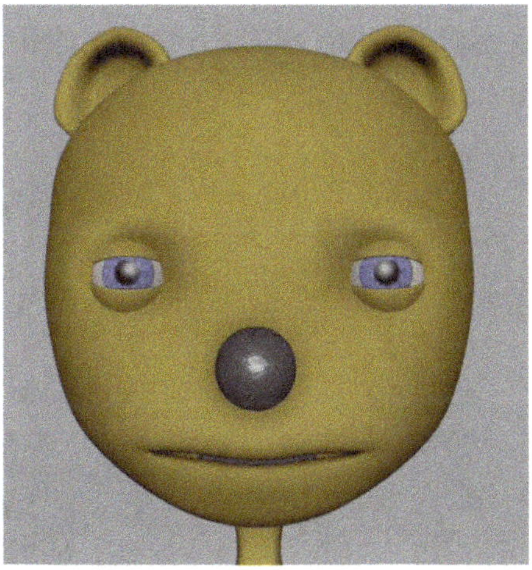

2.5 3D PAINTING

At this point I'll create a temporary *animation texture*—it may not get its final texture until after rigging and animating, depending on the project. Final texture mapping is most conveniently done as part of the lighting and rendering phase of production, so there's no rush on the final texture in most character pipelines.

Popular 3D painting programs include Mudbox, ZBrush, Mari, and Substance Painter. For this simple character, the 3D painting tools built into Maya will suffice.

Rt-click over your *bodyGEO* and choose **Paint > 3D Paint**. You'll get the 3D Paint Tool Settings window popping up, and your cursor changes into a brush. When you hover over your geometry, however, you get a big "no" sign. This is because you first need to have a texture map assigned to paint into. Luckily, the Tool Settings window allows you to create one on the fly.

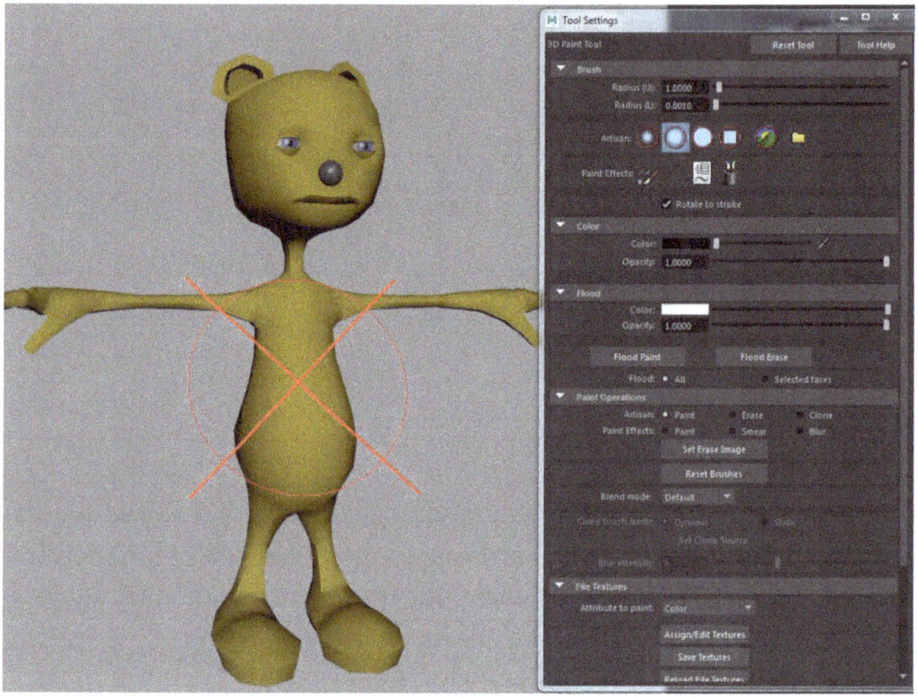

Under **File Textures,** hit the button marked **Assign/Edit Textures**.

In the popup, change **Size X** and **Y** to 1024 and **Image Format** to PNG. Then **Assign/Edit Textures**.

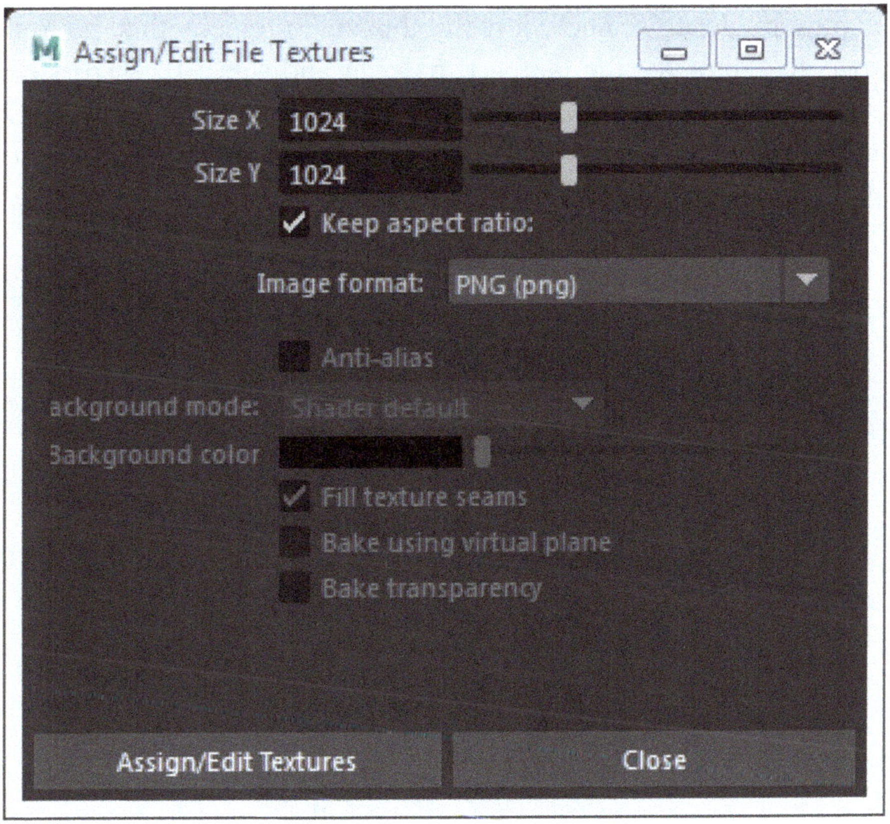

This saves a 1024 PNG file in your Maya project's *sourceimages* directory under *3dPaintTextures*. It will be saved to whenever you hit the **Save Textures** button in the 3D Paint Tool Settings window.

With Maya's color management system on (the default) it is converting your texture file to sRGB space and then converting it again through the View Transform (the on/off button on the panel toolbar). The result is that your image may look too dark in the viewport (do a software render to see how it really looks). To remedy this, go to the Attribute Editor for the lambert material and select the arrow next to the color attribute to navigate to the *file1* node's attributes. Under **Color Space** change it from **sRGB** to **Raw**. Your viewport view should now match a Maya Software render.

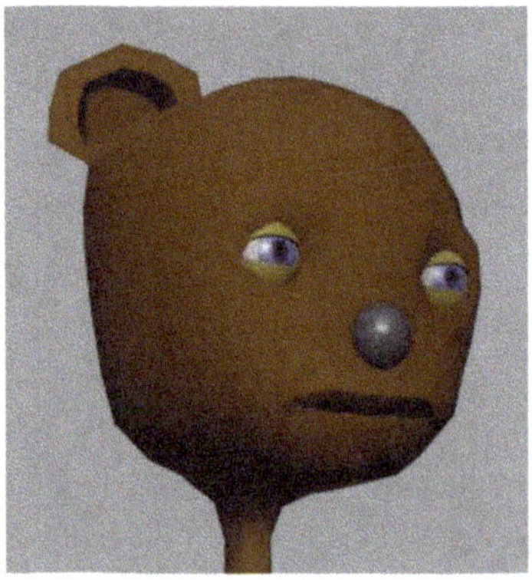

sRGB

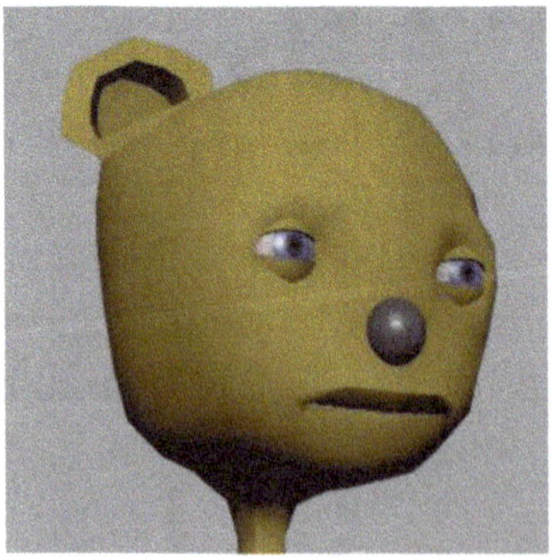

Raw

Now it's time to paint. If you have a pressure-sensitive device (like a Wacom tablet) Maya's paint tools are configured to adjust opacity with pressure. Adjust maximum brush size on the fly by holding the **B** key and dragging left and right. Turn on symmetry under **Stroke > Reflection X**. Paint his lil ol' belly white (a common camouflage trait in the animal kingdom; see *countershading* in Wikipedia). **Save Textures**.

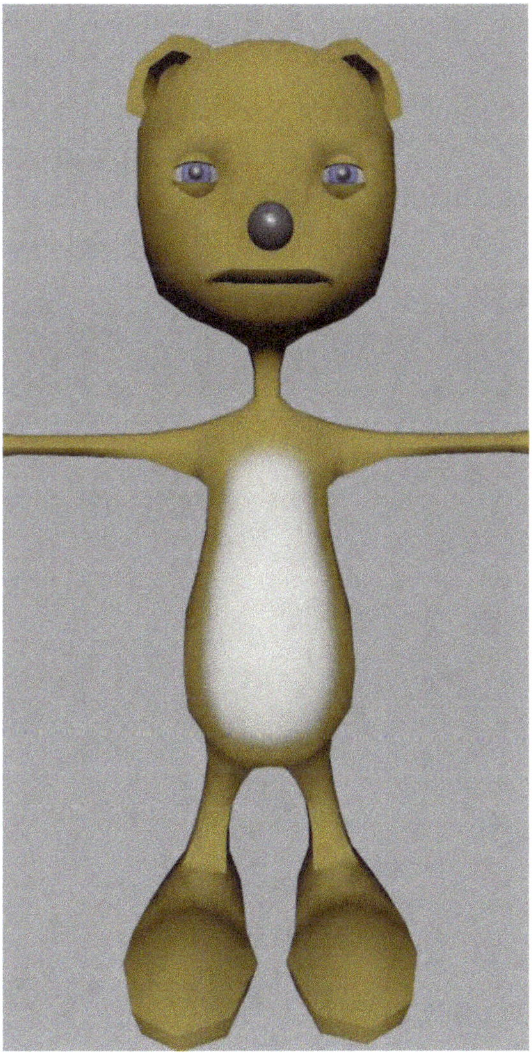

Now paint any other distinctive markings you want and save again. Maybe he needs stripes or polka dots. Be sure to test some of the **Paint Effects** brushes; the watercolors are really cool.

TIP

Tool Settings > Stroke > Screen Projection is good at hiding texture seams. **Paint Operations > Blur** is good at softening artifacts at texture seams.

I just painted some subtle stripes with a paint effects watercolor brush (accessed through the **Get Brush** icon in the 3D Paint tool settings). You can paint in smooth preview mode (**3**) or in standard mode with normals softened (**Mesh Display > Soften Edge**).

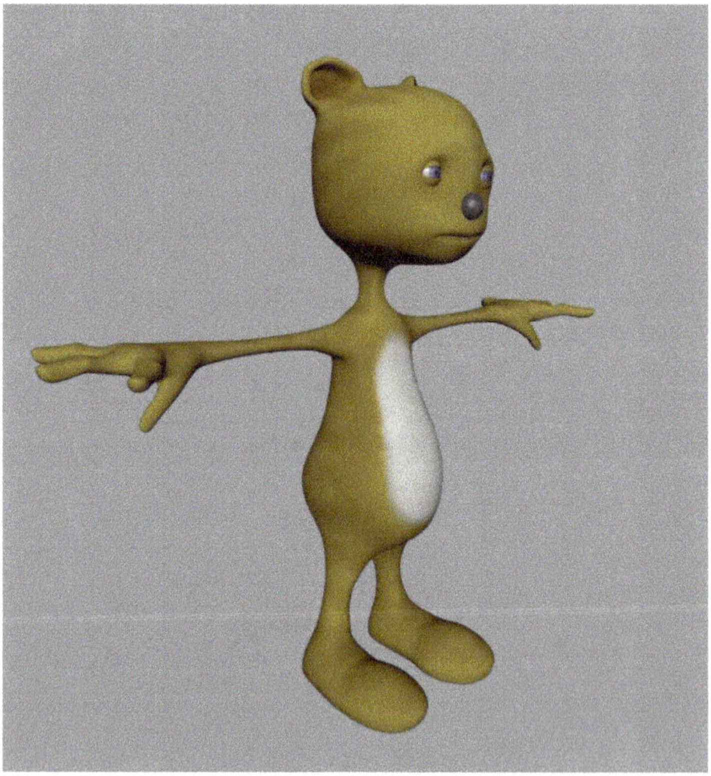

When you are done painting we'll be ready to rig!

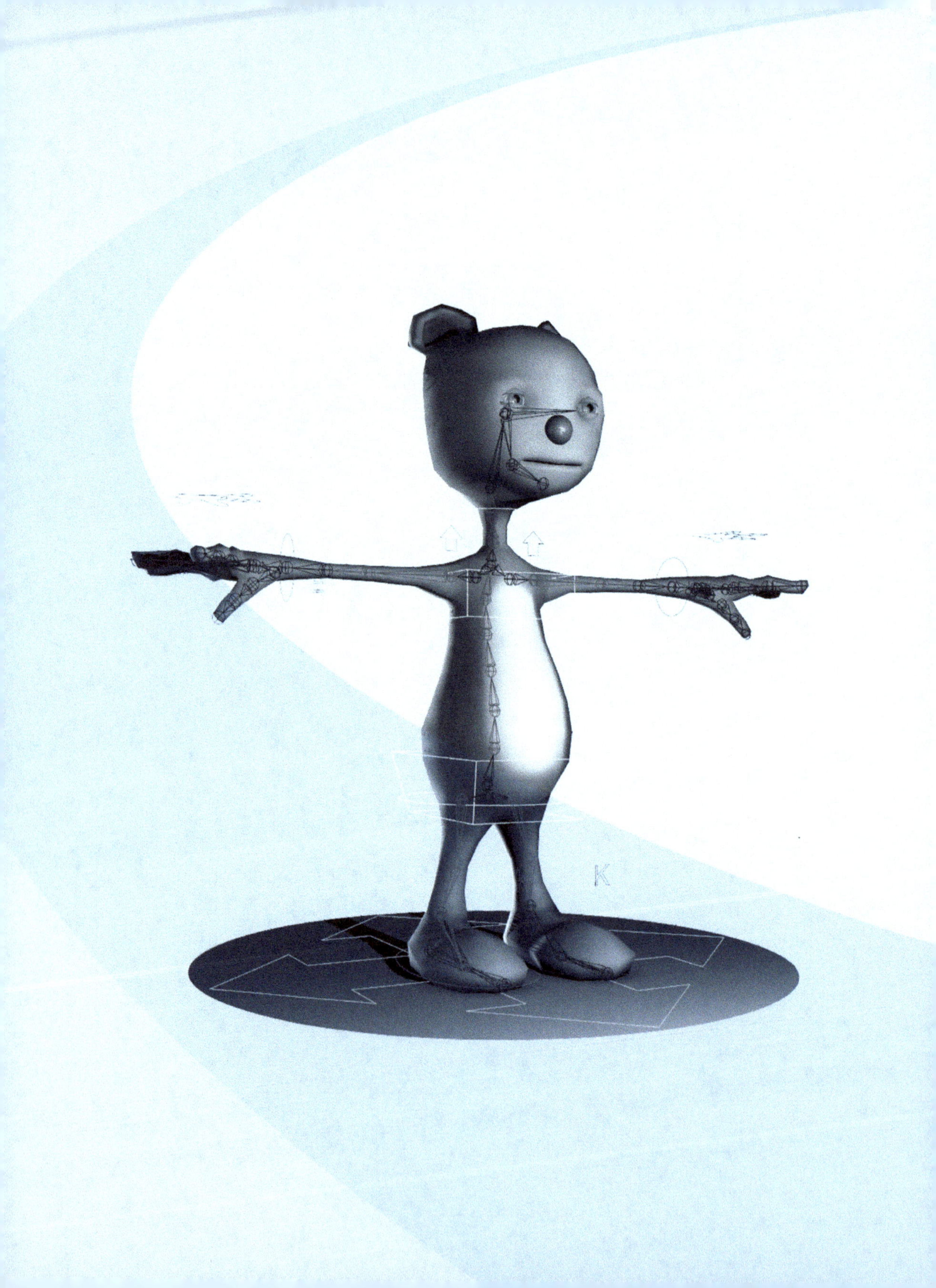

Part III
Character Rigging

At this point, make sure all your geometry and material nodes are named per your naming convention, and group all geometry to keep it sorted from the rigging clutter you're about to make. I selected all meshes in the Outliner, then created a group (Ctrl+G) and named it *geoGRP.*

Make sure all modeling history is deleted and freeze transformations on all meshes (except the eyelids if you want to keep their "zeroed out" state fully closed—I keep mine open while setting up the character but *rotateX* 0 is closed). **File > Optimize Scene Size** to delete any junk nodes left over from modeling.

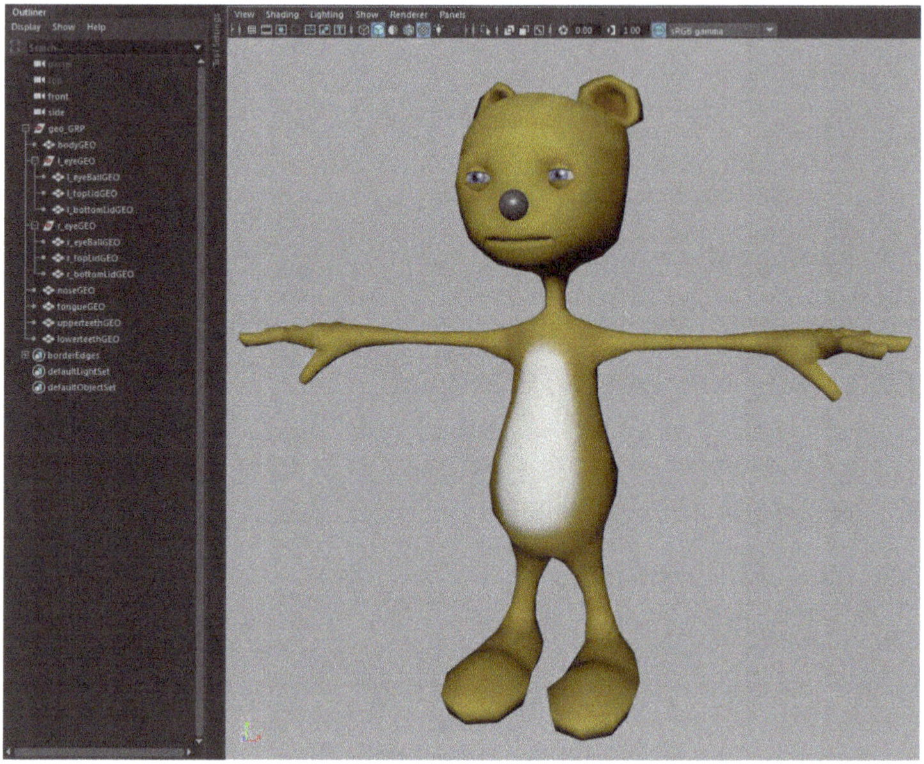

3.1 RIGGING THE LEGS

Remember the point of rigging is to anticipate all of the articulation and range of motion in animation. The best riggers continuously visualize the extremes of movement the character will eventually exhibit throughout the rigging process and will refer to animation motion reference before beginning. YouTube and Vimeo have lots of reference for human and animal locomotion. Take time to plan before you begin.

Put your geometry on a template layer for now. You will only need it as a reference while you lay out the joints for your rig.

Switch to the side view and the **Rigging** menu set. Choose **Skeleton > Create Joints** and click to place 5 joints, as below. Middle mouse moves last joint placed. **Enter** to complete tool. At our scale, joints may appear huge at first. If so, go to **Display > Animation > Joint Size**... and set it to something like .2 so it fits inside the geometry.

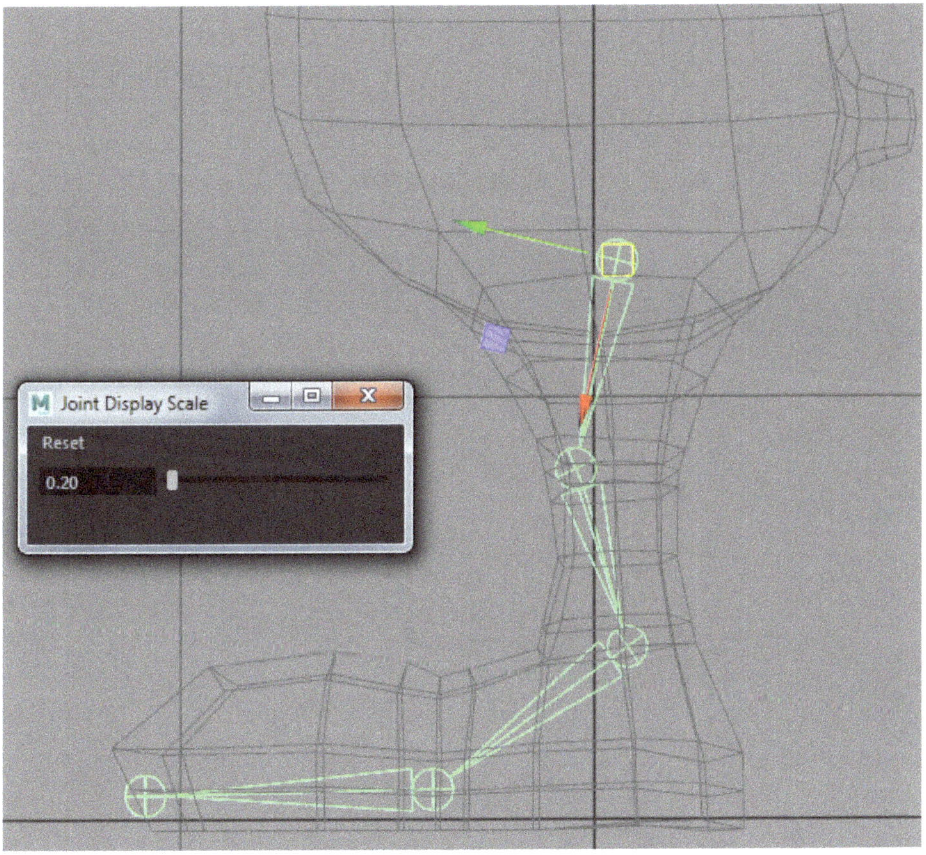

Name the joints ***hip***, ***knee***, ***ankle***, ***ball***, ***toe***. Then select the hip and **Modify > Prefix Hierarchy Names...** with ***left_***

Placement of joints establish two things: the center of rotation for child joints, as well as the center of influence for the skin weights we'll apply later. The former takes precedent over the latter. Place your joints where you think rotation should center from; its correlation is a real-world skeleton, so look at an anatomy reference if you must. The knee joint is favored toward the front of the character because the "bony" side of the joint is usually placed closer to the surface and because it also establishes a "preferred angle" of rotation that will come into play when we add IK solvers later.

In the front view, position the joint chain to fit inside the left leg geometry.

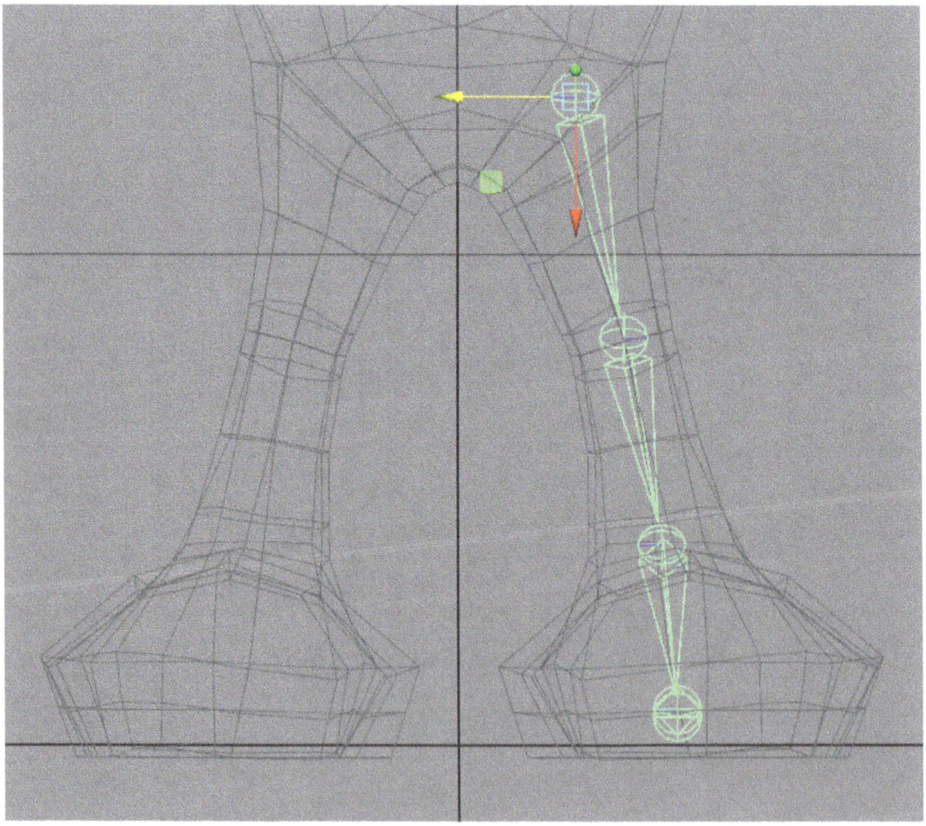

TIP

Up/down arrow keys pick-walk joints, and the **D** or **Insert** keys allow you to move parent joints without moving child joints.

Go to **Skeleton > Create IK Handle (options)** and make sure **Current solver** is set to **Rotate-Plane Solver** and turn on **Sticky**. Then click the hip joint and then the ankle joint. Name this new IK handle ***left_ankleIK***.

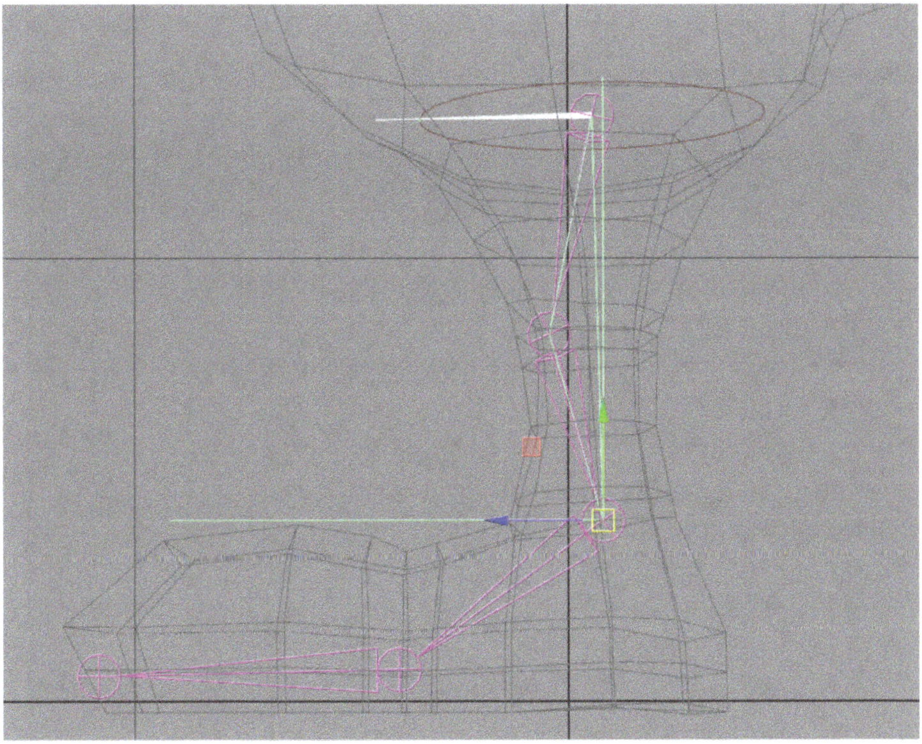

Now go back to **Create IK Handle (options)**, change the **Current solver** to **Single-Chain Solver**, and turn on **Sticky**. Create an IK handle from the ankle to the ball joint and one from the ball to the toe joint. Name these ***left_ballIK*** and ***left_toeIK***.

TIP

IK handle size can be controlled similarly to joint size, with **Display > Animation > IK Handle Size...**

Select the top bone in the hierarchy, go to **Skeleton > Mirror Joint (options)** and mirror across **YZ** with function set to **Behavior**. Also, under replacement names, search for ***left_*** and replace with ***right_***.

This will duplicate your leg system across to the right side.

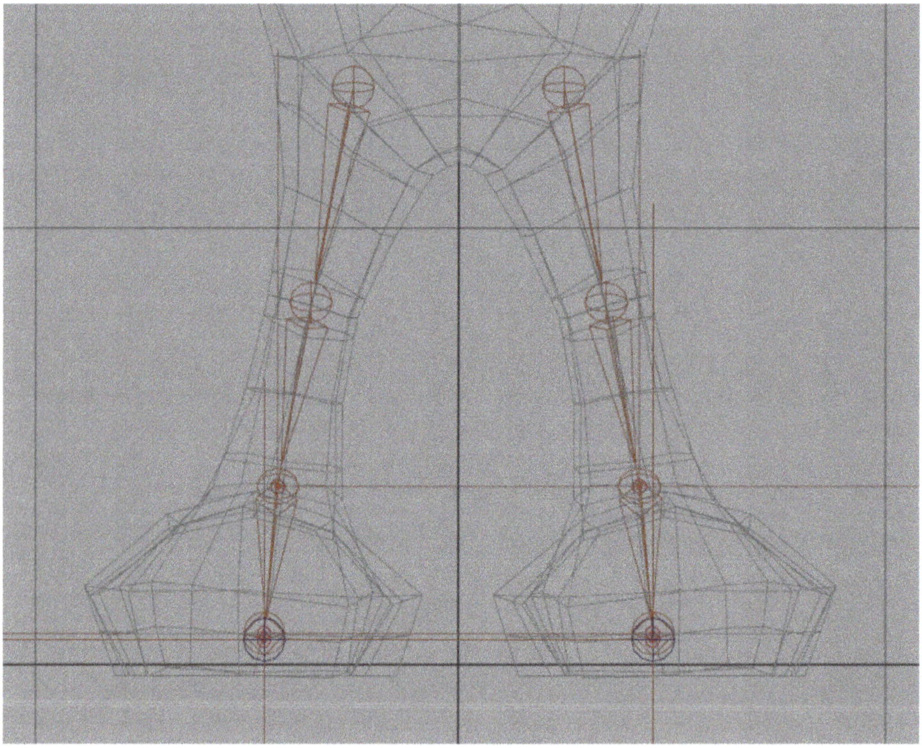

3.2 RIGGING THE FEET

Now we'll set up the feet with a simple reverse-foot setup. Starting at the character's heel, place a joint in the side view and then hold **V** (point-snap) to place joints at the toe, then ball, then ankle, going in the reverse direction from your original chain. Name these joints ***revHeel***, ***revToe***, ***revBall***, ***revAnkle***.

TIP

To accurately place the first (heel) joint without leaving the side view, you can create it in an empty area, then point-snap it (**V** with middle mouse button) to the ball joint, and then middle-mouse-drag it behind to the heel.

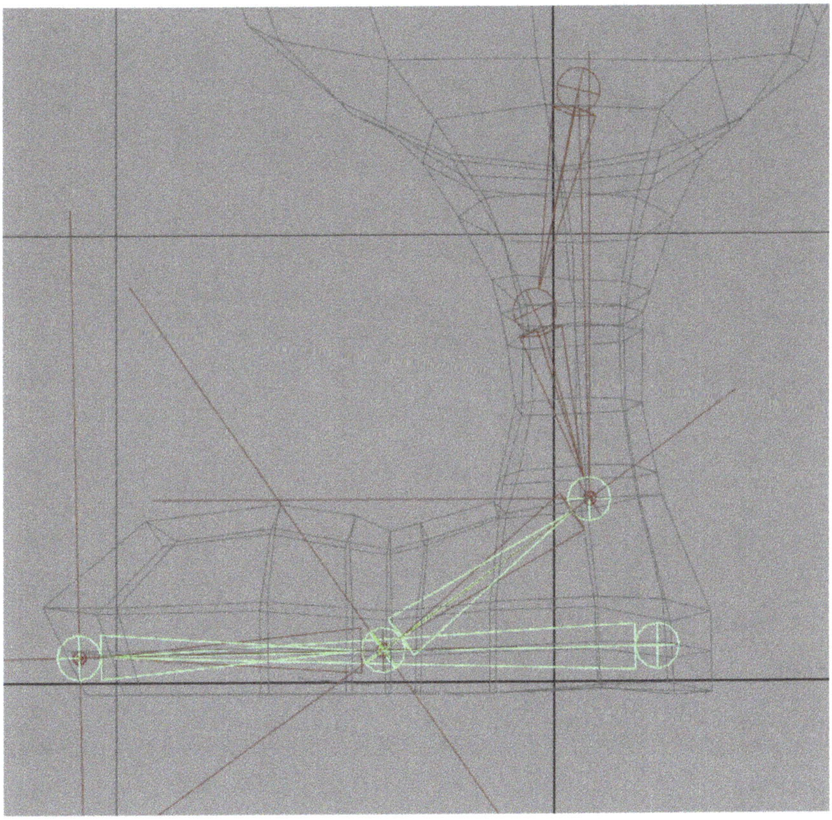

Prefix hierarchy names with ***left_*** and then **Skeleton > Mirror Joints** on this reverse chain, replacing ***left_*** with ***right_***.

Now parent the leg's IK handles under the corresponding joint in the reverse foot (select the handle, than Ctrl-select the reverse foot joint in the Outliner and hit **P** to parent)—***ankleIK*** to ***revAnkle***, ***ballIK*** to ***revBall***, ***toeIK*** to ***revToe***. You can test this setup by lifting and rotating the ***revHeel*** bone. Undo after testing to reset foot position. Remember to do this for both feet.

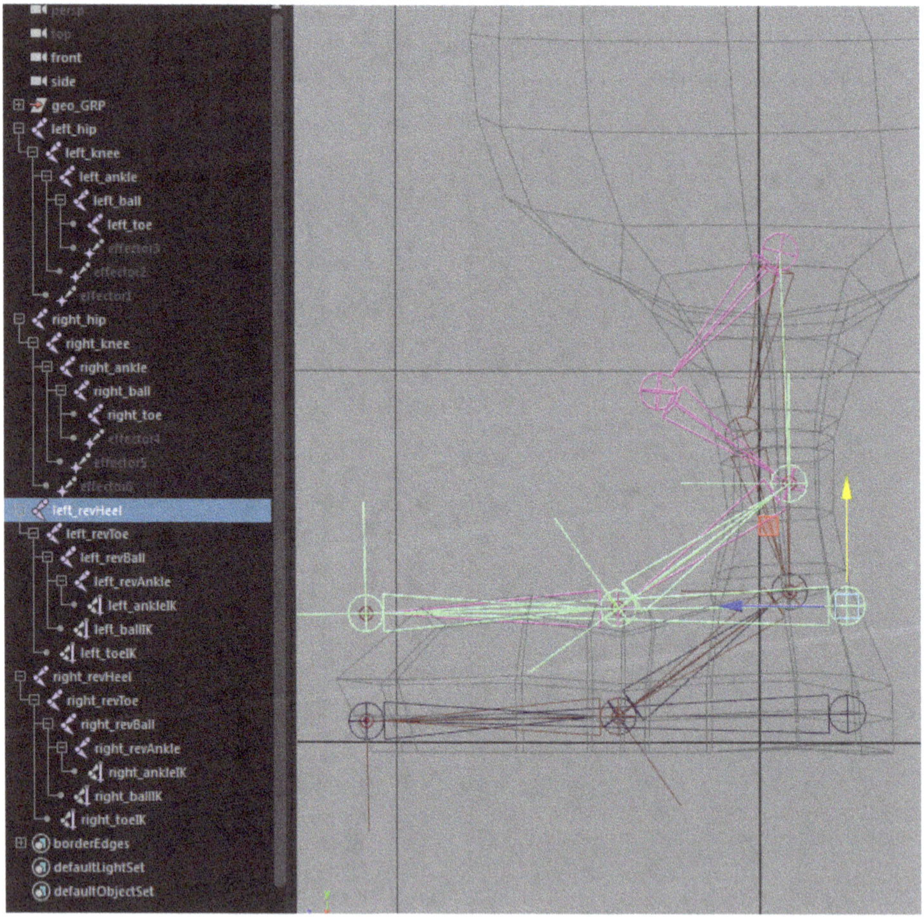

Next we'll create a foot control object. Character setup involves creating a set of easy-to-select manipulators for animators to set keys on, thereby animating parts of the underlying rig. We will try to make all of our control objects out of NURBS curves, because they don't render by default. **Create > NURBS Primitives > Circle** and with the move tool, **V** + MMB-click on the ball joint to snap the center of the circle around the ball (**V** = point snap). Scale it to fit around the templated foot geometry.

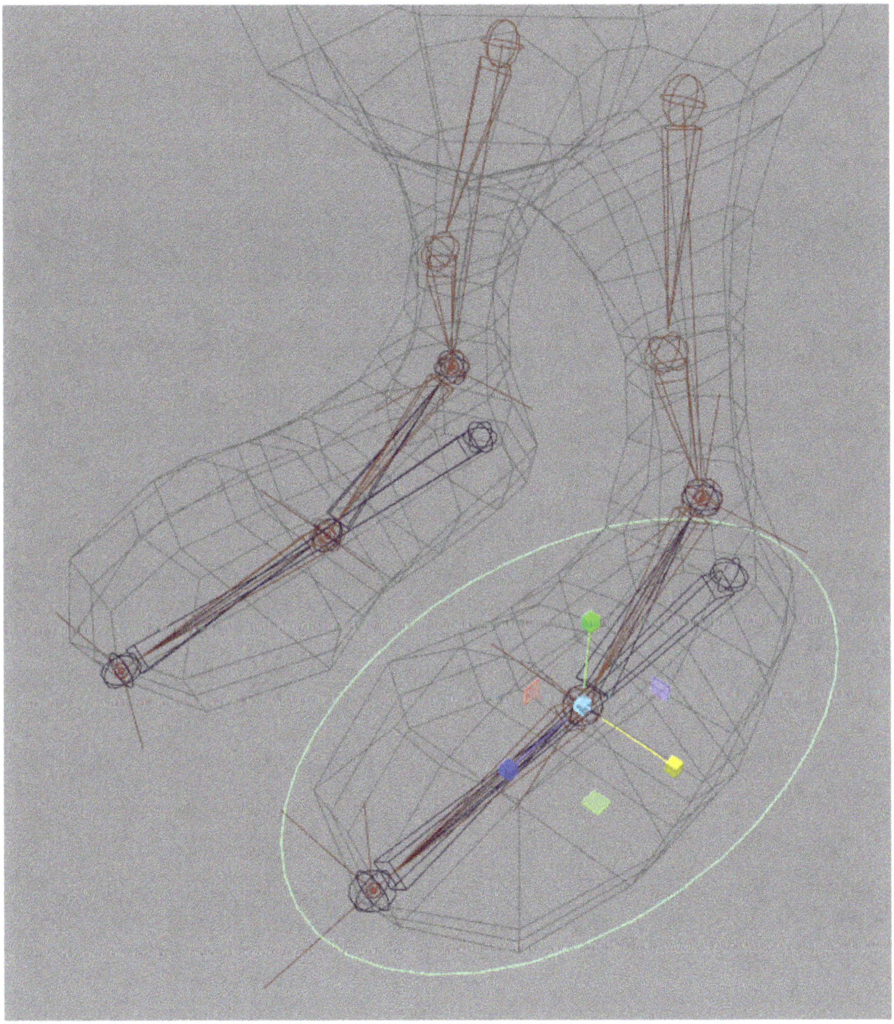

Switch to control vertex mode (**F8** or RMB over the curve to select CV mode) and fashion the CVs of the curve to a shape more approximating the foot. Animation control curves should be outside the geometry so they can be easily selectable in shaded mode, but placed intuitively close to the objects/parts they control.

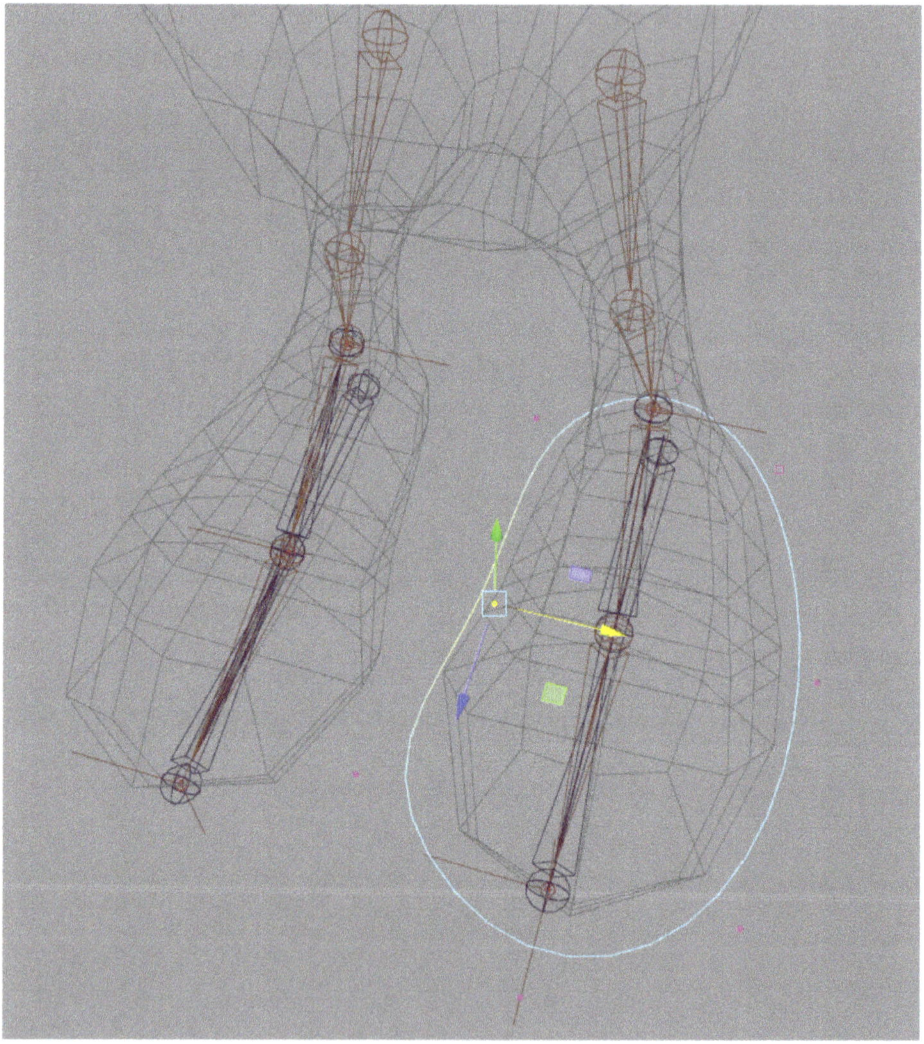

Name this control object ***left_footControl***. Duplicate a copy flipped negative in **scaleX**, and then add a negative to the **translateX** value to pop it over to the right side. Rename this copy ***right_foot-Control***. Freeze transformations and delete history on both, and then move the pivot points to their respective heel joints (hold **D** and **V** while MMB-snapping the pivot point to the *revHeel* bone).

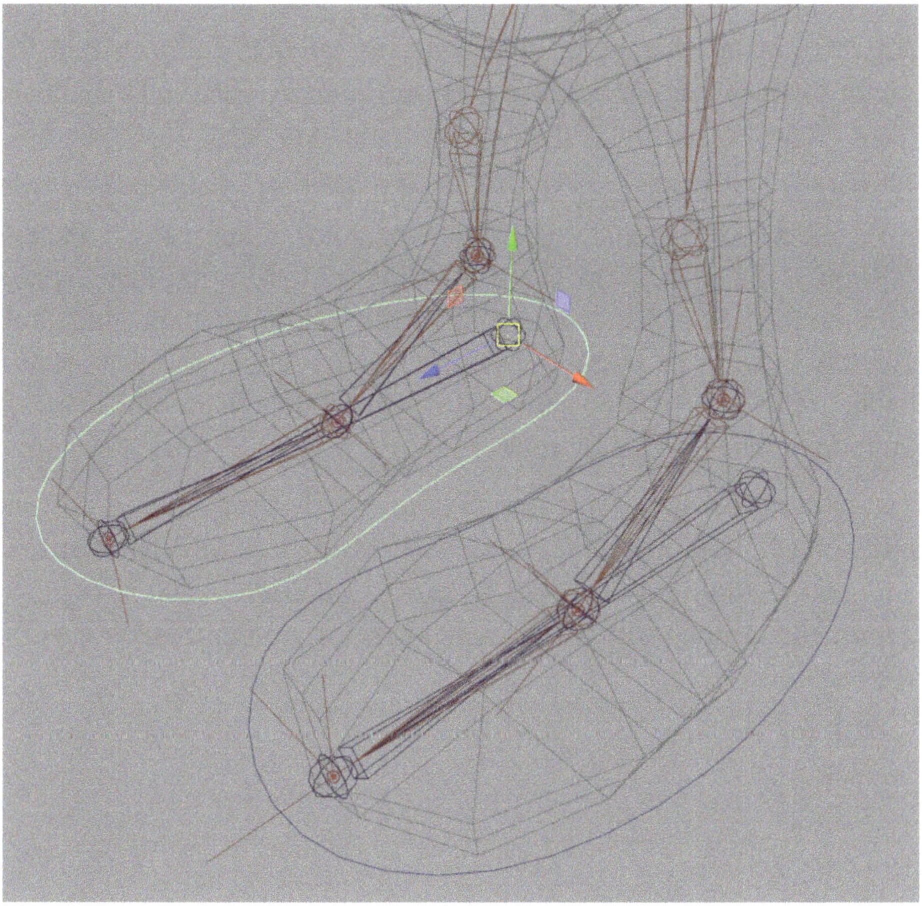

Both foot controllers currently have the same orientation, but not mirrored behavior. In other words, when selected together, they will rotate in the same direction, like a skier. Animators find it more intuitive to have opposing movements in a pair of counterpart limbs; for example, to rotate the feet to point symmetrically inward or outward with a single manipulation. To do this, it takes a few more steps.

Parent *footControl* under the *revHeel* joint, and then group it (Ctrl+G). Name this group ***left_footControlGRP*** and unparent the group (Shift+P). Freeze transforms on the *footControl*. This causes the *footControl* to inherit the local rotation axis of the *revHeel* joint.

Parent constrain each *revHeel* joint to its respective *footControl* (select *footControl*, shift-select *heel* joint, **Constrain > Parent (options)**. Make sure **Maintain offset** is set to **On** and then **Apply**. This establishes a parent-like relationship while keeping joints and controls in separate hierarchies. Now you can move the feet around by their *footControl* nodes, and when selected together with the rotate tool, their behavior should mirror.

Shift-select both *footControl* objects, then add some custom attributes to help drive the animation. **Modify > Add Attribute...** and in the dialog, set **Long Name** to ***ballRotZ***, **Data Type** to ***Float***, then click **Add**. Add all of these attributes before closing the window:

ballRotZ
toeRotX
toeRotY
toeRotZ

Open **Windows > General Editors > Connection Editor**. Select one of the *footControl* objects and click the **Reload Left** button, then select the corresponding *revBall* joint and hit **Reload Right**. In the left column, select the **ballRotZ** attribute and in the right column, open the **Rotate** section (currently greyed out) and select the **rotateZ** attribute. This creates a direct connection between the two attributes.

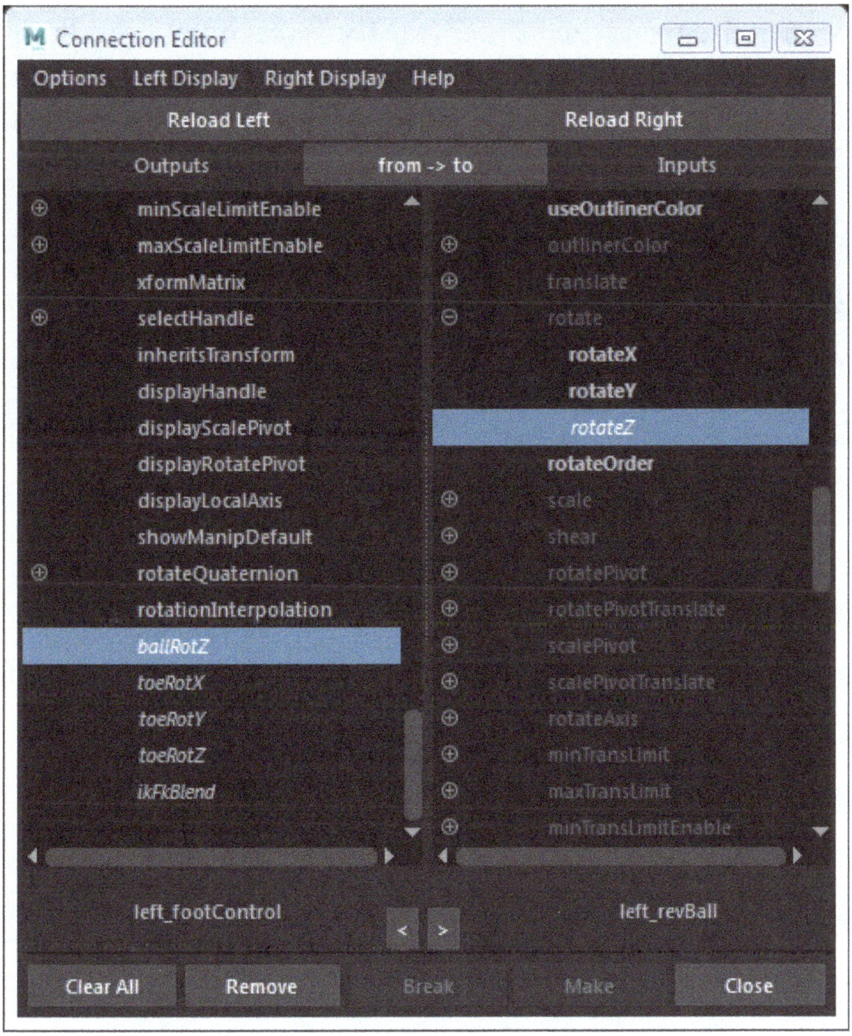

Make the following connections:

footControl **ballRotZ** to *revBall* **rotateZ**

footControl **toeRotX** to *revToe* **rotateX**

footControl **toeRotY** to *revToe* **rotateY**

footControl **toeRotZ** to *revToe* **rotateZ**

Now test all the custom attributes on each *footControl*. Select a *footControl*, highlight the name of an attribute in the **Channel Box**, then use the virtual slider (MMB dragged left or right in the viewport) to slide through values to see if the proper joints move. If a wrong connection was made, go back to the **Connection Editor**, **Reload Left** and **Right** again, and deselect the connected attributes to break the connection. Set values back to **0** when done testing.

TIP

If you want to add limits to each control, you can do so by selecting a *footControl* and **Modify > Edit Attribute...** Turn on **Has Minimum** or **Has Maximum** and type in a value based on your testing. Optional.

Select both *footControls* again and **Modify > Add Attribute...** Create a new attribute, ***ikFkBlend*, Data Type *Float*, Minimum 0, Maximum 1, Default 1**. With the **Connection Editor**, connect each *ikFkBlend* attribute to the *ikBlend* attributes of all three IK handles on the respective leg.

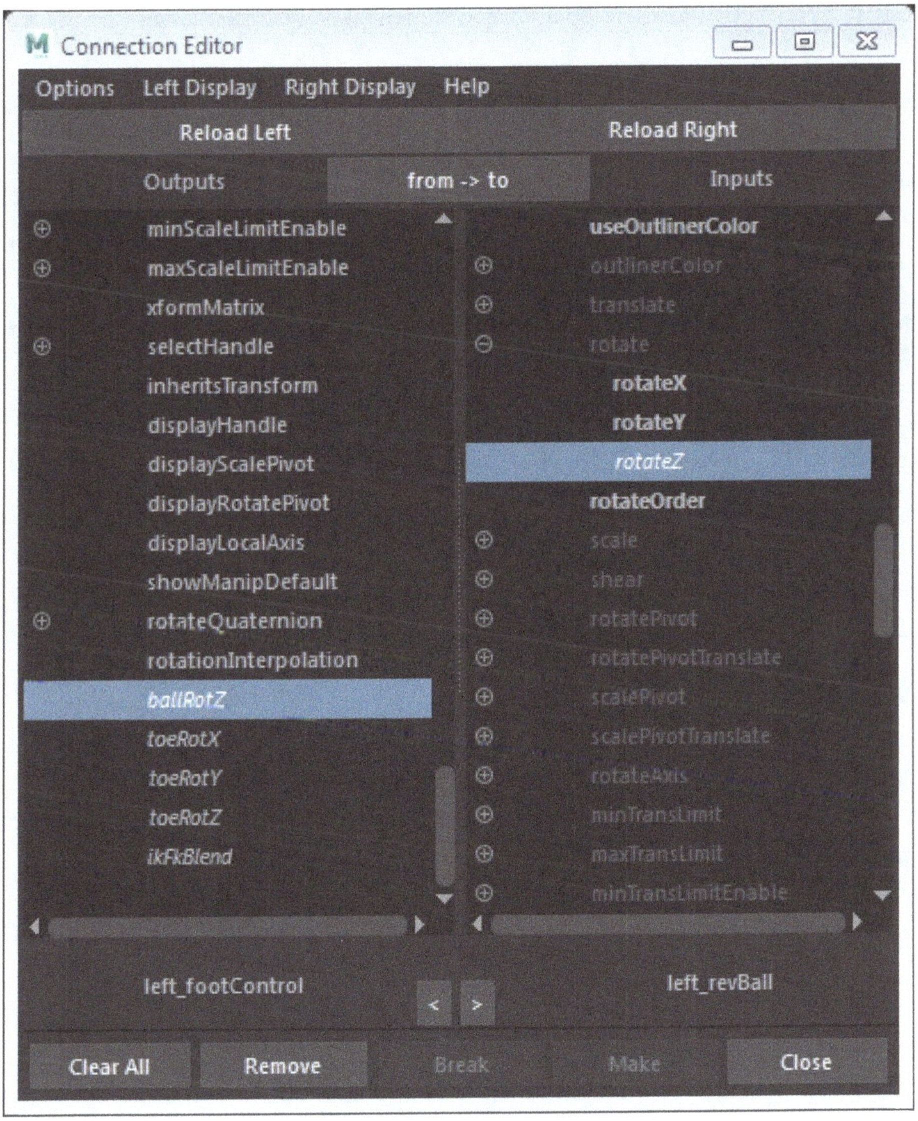

Now you can use the *IkFkBlend* attributes on the *footControls* to disable all IK when you want to animate the legs with FK (for example, when the character is hanging from something with legs swaying).

To clean up each *footControl*, select **Scale X, Y, Z** and **Visibility** in the channel box and RMB over the attribute names to **Lock and Hide Selected**.

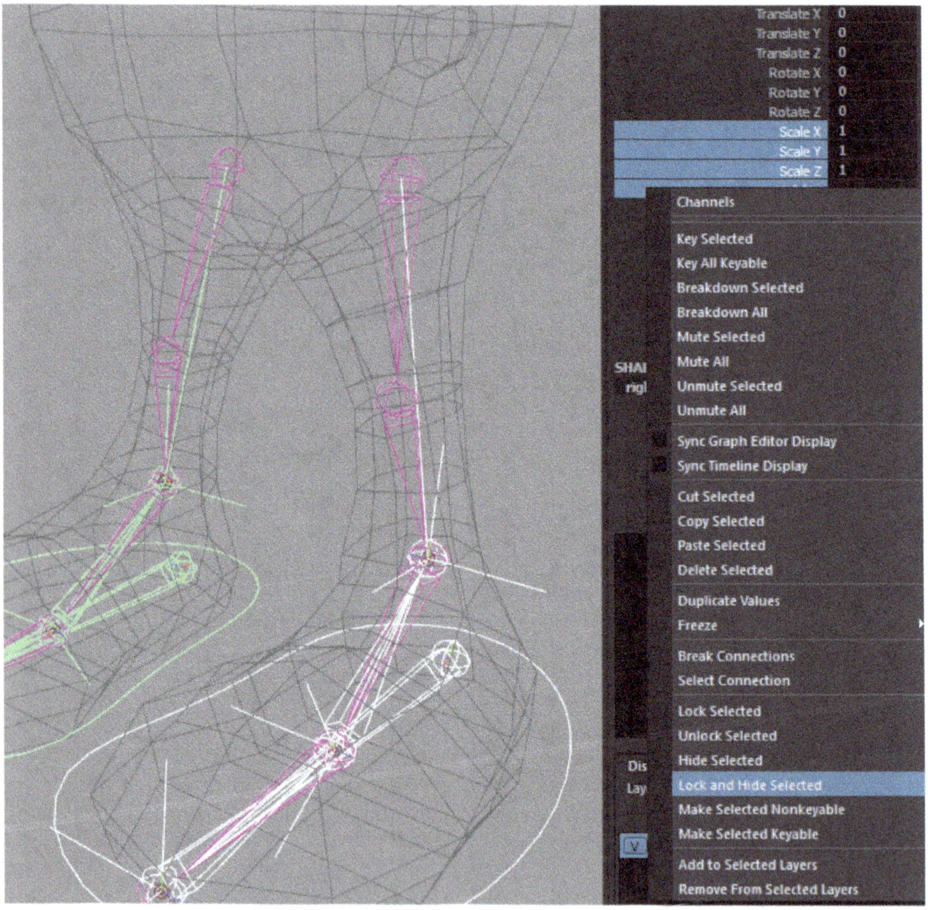

Now we'll create Pole Vector manipulators to control the direction of the knees. Text curves make good manipulators for this. **Create > Type.** In the **Attribute Editor > type1** tab, enter the text "K" and adjust the font if desired. Under the **Geometry** tab > **Mesh Settings > Create Curves from Type.** Now you can delete the polygon type and keep the NURBS curve to use as a control object.

Modify > Center Pivot and then scale the "K" icon to a good size. Snap it to the knee joint and then pull it forward in Z so that it is in front of the leg at knee level. Ctrl+D to duplicate it and snap the new one to the other knee joint and pull forward to the same translateZ value. Name them ***left_kneeControl*** and ***right_kneeControl***. Freeze transformations and delete history.

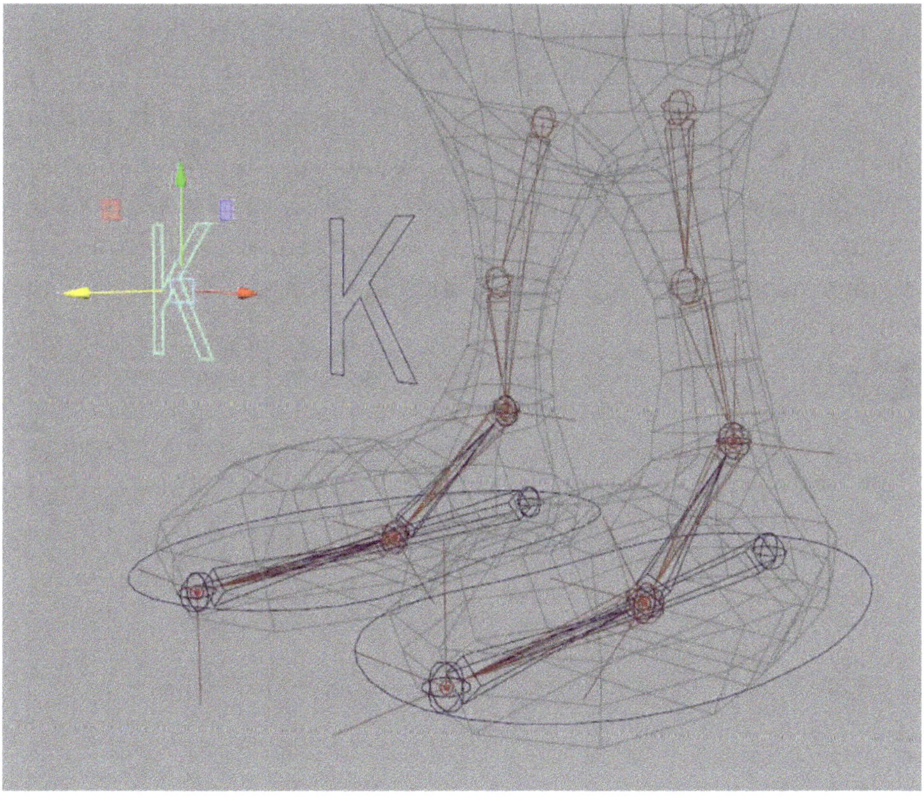

Select ***left_kneeControl***, shift-select ***left_ankleIK*** and **Constrain > Pole Vector**. Do the same for the other leg. Test by moving the *kneeControls* and witness the knees rotating. Lock and Hide all attributes except for **Translate** on the *kneeControls*.

Select the IK handles and set their **Visibility** to **off** to keep them from being accidentally selected.

3.3 RIGGING THE ARMS

Go to the top view and place joints for the arm according to the picture below. Place the first joint halfway between the neck and shoulder—name it ***clavicle***; next one in the center of the shoulder—name it ***shoulder***; third one ***elbow***, fourth one ***wrist***. Continue to create a joint in the middle of the palm (***hand***) and then do the fingers (***index1, index2, index3, middle1, middle2, middle3, pinky1, pinky2, pinky3***). Switch to the front view to lift the chain off the grid and place the bones inside the arm geometry. Click on the ***hand*** joint to start the chain for the thumb: ***thumb1, thumb2, thumb3.*** Note that we are leaving out the third knuckle on each finger to keep with the cartoon-hand convention—a realistic character would have another joint in each finger.

TIP

In the input field of the Status Line there's a drop-down that switches it to **Rename** mode—you can use this to rename multiple objects at once and they will automatically be given index numbers.

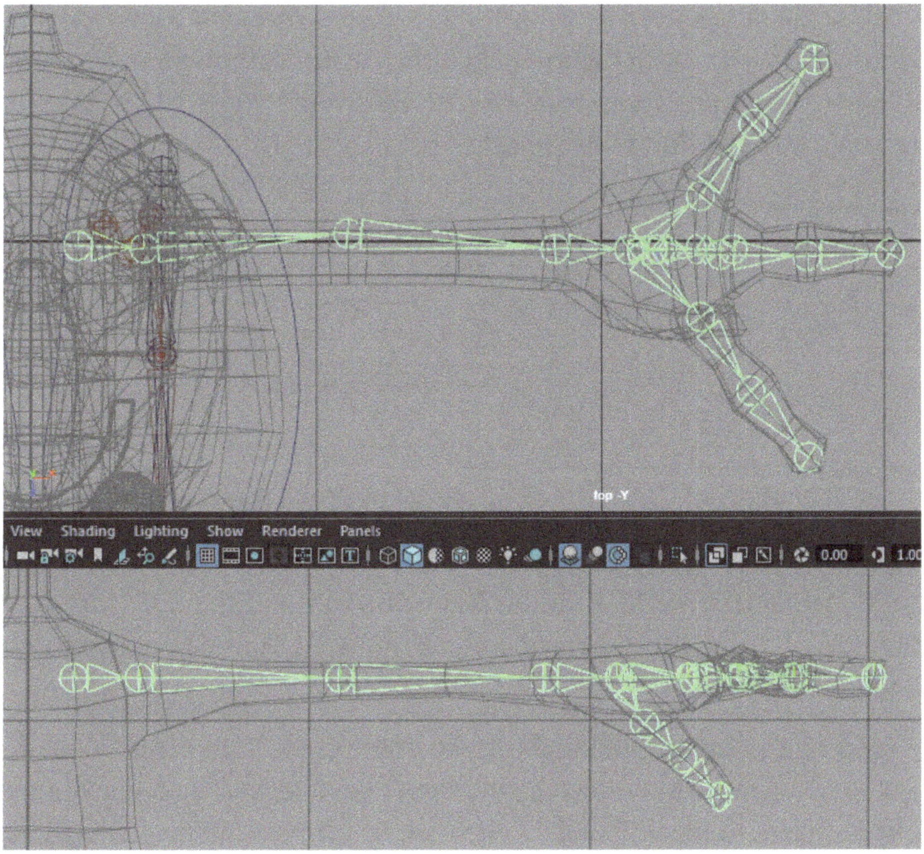

TIP

To continue placing joints from an existing joint, just click on the joint first with the joint tool before adding new ones. To navigate up/down the hierarchy while placing joints, use arrow keys. **Backspace** while placing joints deletes the most recently placed joint. **Insert** key with the move tool moves a joint without moving its children. Always move joints into place (rather than rotating or scaling) for best practices—by keeping rotational values clean you can always "zero out" a joint later to return it to its default position, and it allows you to auto-correct rotation axes with **Skeleton > Orient Joint**.

Use **Skeleton > Insert Joints** to create a ***forearm*** joint between elbow and wrist. With the tool, click the existing elbow joint and drag the new joint to halfway between the elbow and wrist. (This ensures it is in a straight line with the other bones.)

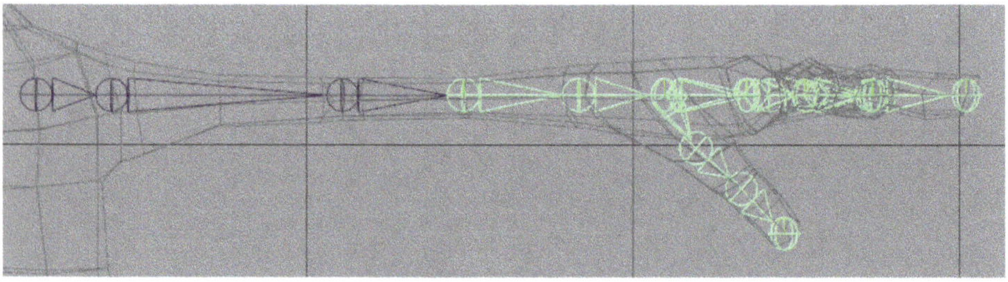

Select the ***clavicle*** and **Modify > Prefix Hierarchy Names…** with ***left_***.

Create ***left_armIK*** from the ***left_shoulder*** and the ***left_forearm***. Set the solver to **Rotate-Plane Solver** in the IK Handle Tool settings, and set **Stickiness** to **off**.

Go to the Outliner and find the end effector under the elbow joint; select it and, with the Move tool, hit the **Insert** key and hold **V** to snap the end effector pivot to the wrist joint. Then toggle off pivot mode (**Insert**). Go into the Attribute Editor for the *left_armIK* handle and under **IK Handle Attributes**, set **Stickiness** to **sticky**. (The reason we create a forearm joint is to divide the rotation of the forearm between two joints to minimize distortion of the skin. We'll create a utility node to handle this later.)

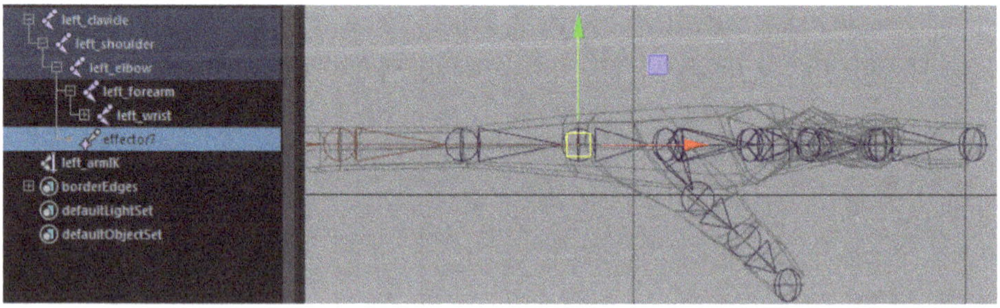

Create > NURBS Primitives > Circle and hold **V** to snap it to the wrist bone. Rotate in 90 in Z and scale it so it fits outside the surface of the character's wrist. Name it ***left_armControl***.

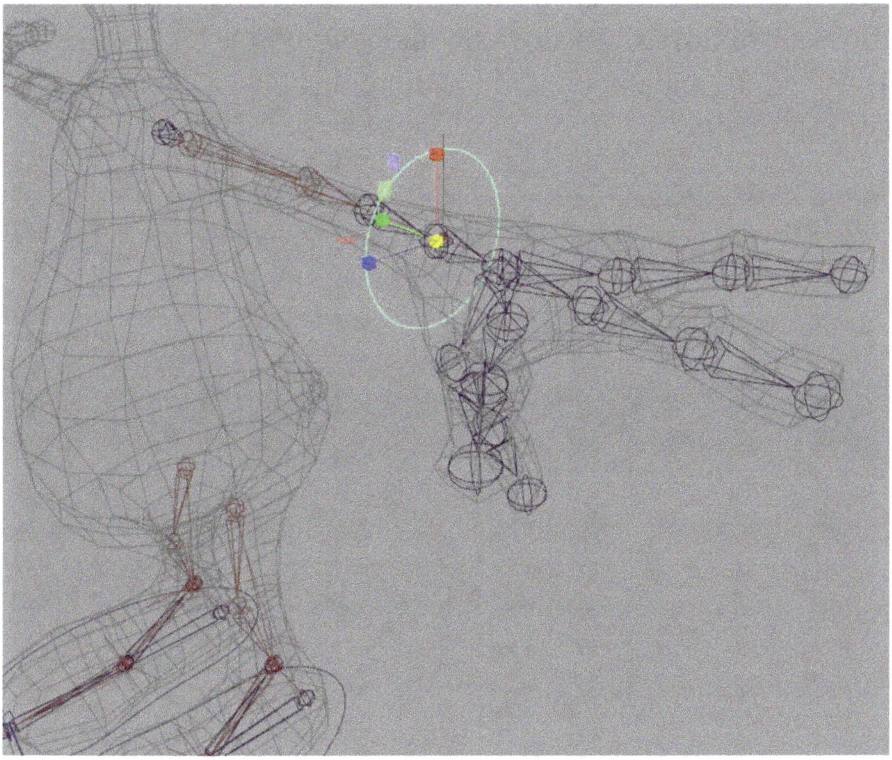

Parent *left_armControl* under the *left_wrist* joint, and then group it (Ctrl+G). Name this group ***left_armControlGRP*** and unparent the group (Shift+P). Freeze transforms on the *left_armControl*. This causes the *left_armControl* to inherit the local rotation axis of the *left_wrist* joint.

Freeze transformations and delete history on *left_armControl*. Select *left_armControl* and Shift-select *left_armIK* and **Constrain > Point (options)**, making sure **Maintain offset** is **On**. Select *left_armControl* and Shift-select the wrist joint, then **Constrain > Orient** (also with **Maintain offset**). Now the *left_armControl* both moves the hand in IK and rotates it in FK. Lock and hide its **Scale** and **Visibility** attributes.

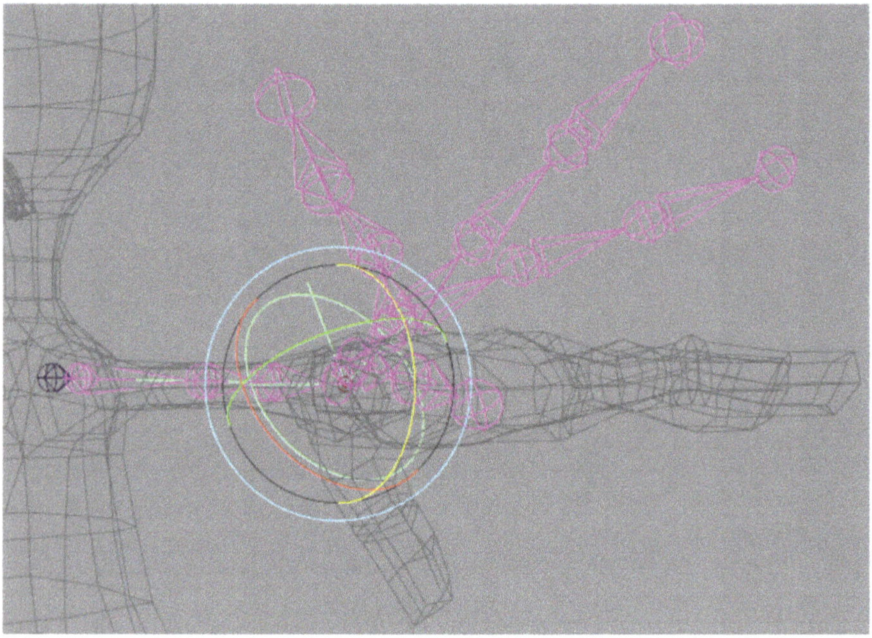

Create > Type and type the letter "E". **Create Curves from Type** similar to how we did the knee controls; center pivot. Scale down and snap this to the elbow and pull back in Z so that it is behind the character's elbow. Freeze transformations and rename it ***left_elbowControl***. Select it, Shift-select the arm IK, and **Constrain > Pole Vector**. Lock and hide its **Rotation**, **Scale**, and **Visibility**.

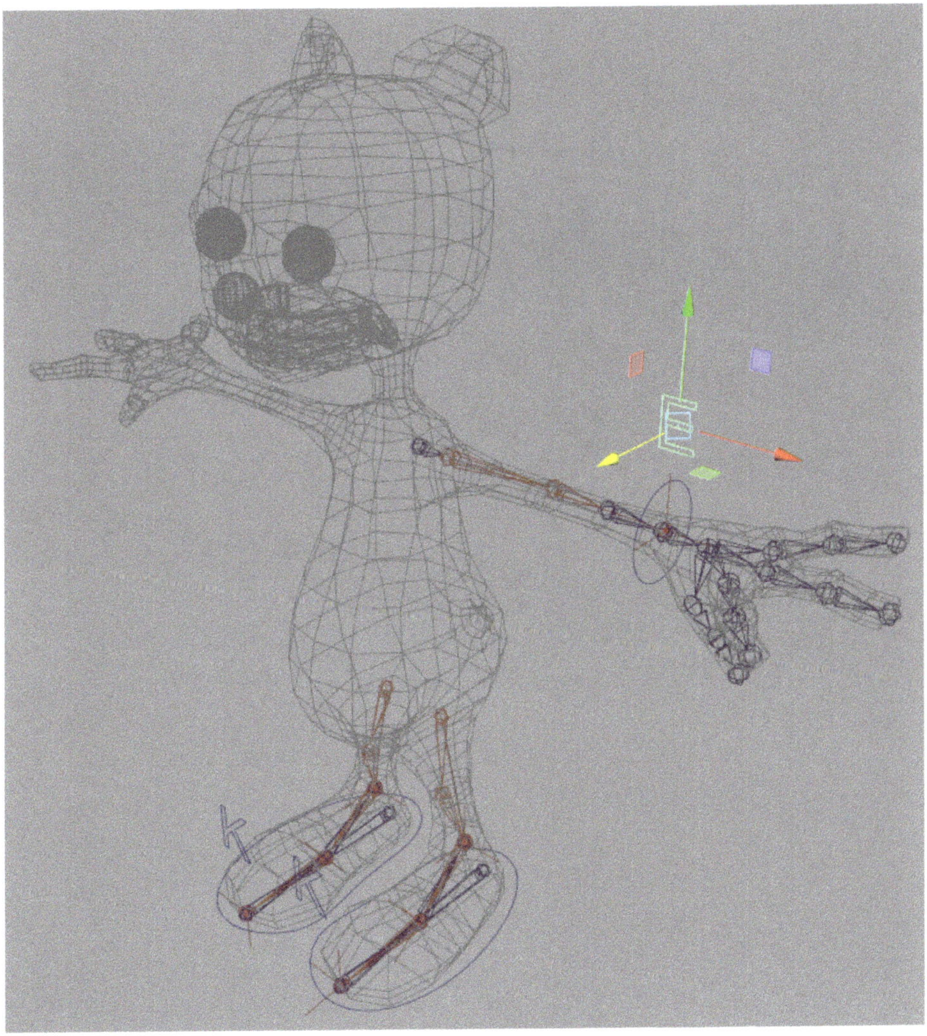

Now to create a clavicle control. **Create > Curve Tool > EP Curve Tool (options)** and set **Curve Degree** to **Linear**. In the front view, hold **X** to snap to grid to draw the following arrow:

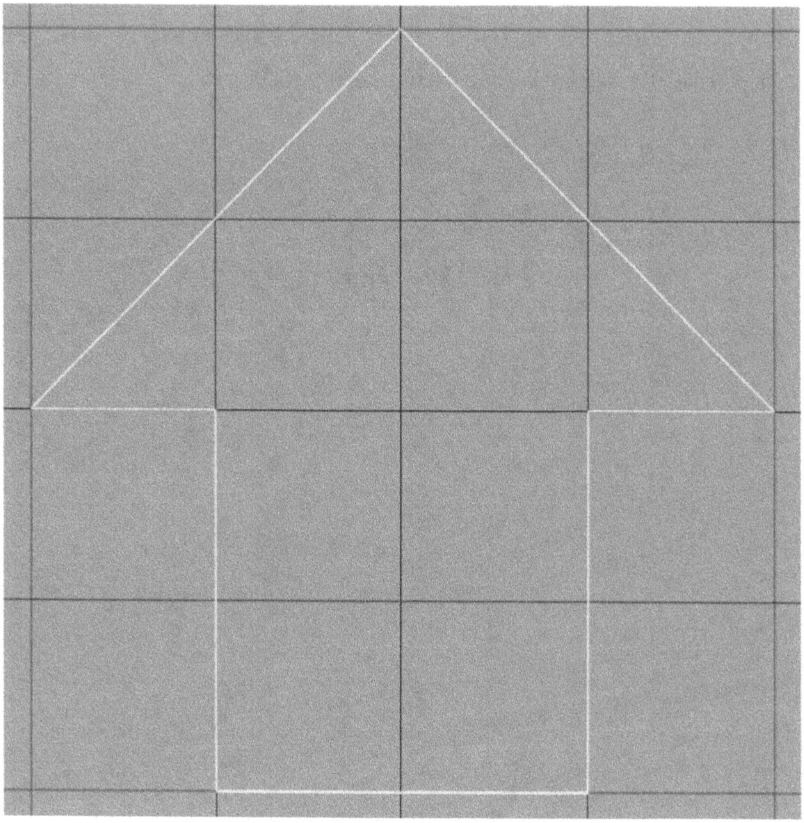

Name it ***left_clavicleControl*** and **Modify > Center Pivot**. Move and scale it so it is above the left shoulder. Snap its pivot to the clavicle joint. Parent it under the *left_clavicle* joint, and then group it (Ctrl+G). Name this group ***left_clavicleGRP*** and unparent the group (Shift+P). Freeze transforms on the *left_clavicleControl*. This causes the *clavicle-Control* to inherit the local rotation axis of the clavicle joint. Now select the *left_clavicleControl*, Shift-select the *clavicle* joint, and **Constrain > Parent (options)**. Set **Maintain offset** to **On** and constrain. Lock and hide *left_clavicleControl*'s **Translate**, **Scale**, and **Visibility**.

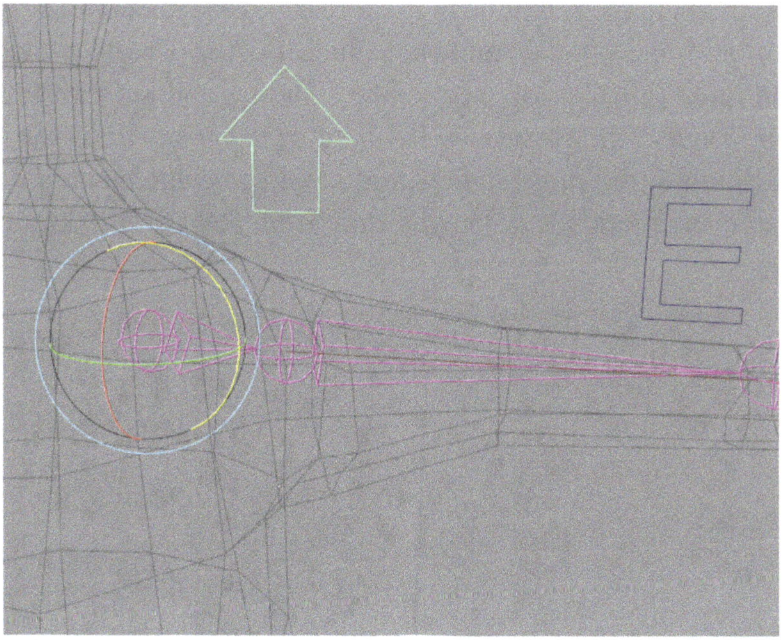

Select the *left_armControl* and **Modify > Add Attribute...**

Long Name: *ikFkBlend*

Data type: Float

Minimum: 0

Maximum: 1

Default: 1

Use the Connection Editor to connect this new attribute to the *ikBlend* of the IK Handle. In the Attribute Editor for *left_armIK*, turn on **IK FK Control** under the **IK Solver Attributes**. Also connect the *ikFkBlend* attribute of the *left_armControl* to the *left_armControl-WO* attribute in the *left_wrist_orientConstraint1* node (look under the wrist joint in the Outliner) —this will enable and disable the orient constraint when switching to FK mode so that you may manually rotate the hand. Now you can select the IK handle and set its visibility to **off** to keep it from being accidentally selected.

To automate the roll bone in the forearm, first select the *left_armControl* and open the **Windows > Node Editor.** Create a **Multiply Divide** node (hit the Tab key in the window and start to type the name). Connect the **Rotate X** attribute of *armControl* to the **Input 1X** attribute of the *multiplyDivide* node. (The *unitConversion* node is created automatically when you drag from **Rotate X** to **Input 1X**.)

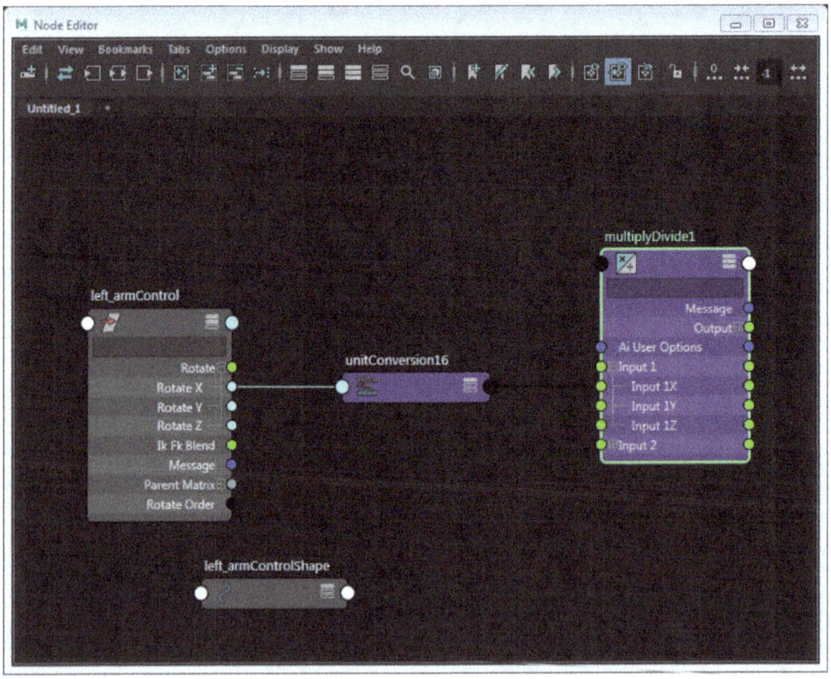

Open the **Attribute Editor** for the *multiplyDivide* node and **set Input 2 X** to **0.5**. In the **Connection Editor**, load the *multiplyDivide* node in the left side and the *left_forearm* joint on the right (or add *left_forearm* to the Node Editor graph). Connect the **Output X** attribute of the *multiplyDivide* node with the **Rotate X** attribute of the forearm joint. Now half of the x rotation of the wrist will go to the roll joint.

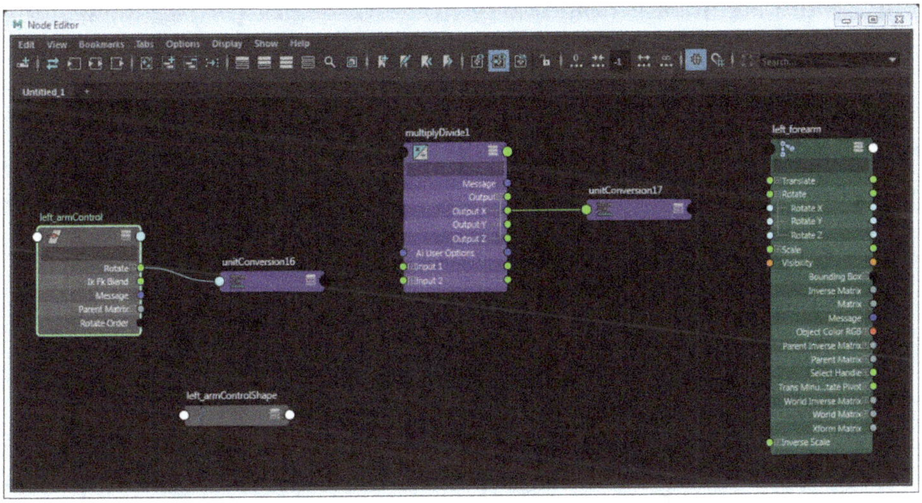

Roll joints allow the rotation of the hand to be shared between two joints so that unsightly skin pinching does not occur when the hand twists a full 180 degrees. It is mostly an issue at the wrist due to its extreme range of rotation, but you can use this same technique to create roll joints anywhere you have noticeable twisting motion.

3.4 RIGGING THE HANDS

NURBS circles get boring when overused for control objects, and they can get confusing when they all look the same. To mix things up, I suggest a shelf of handmade control objects of various shapes, or you can download a free generator, such as Sin's rig controllers (*https://www.highend3d.com/maya/script/rig-controllers-for-maya*). For a more advanced toolbox, try mz_ctrlCreator (*https://www.highend3d.com/maya/script/mz_ctrlcreator-for-maya*).

For the hands, I recommend putting finger FK controls on a separate object, since the IK object will be left behind in FK mode. I just doodled an outline of a hand in the top viewport with **Create > Curve Tools > CV Curve Tool**:

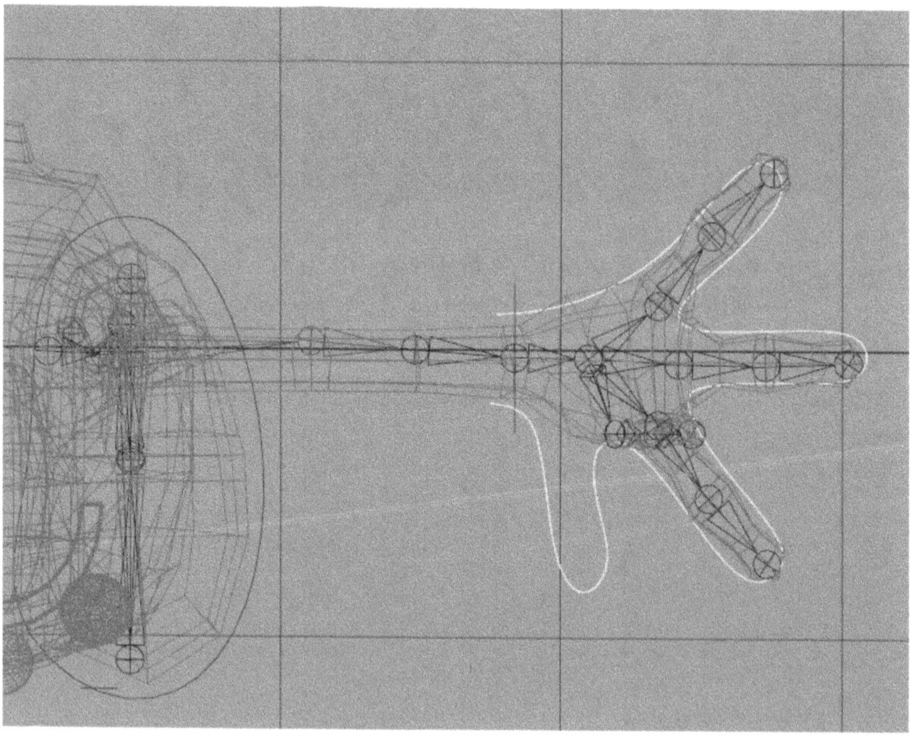

Center the pivot, scale it down to not be so big, place it just above or behind the hand and parent constrain it to the hand joint (select *left_hand*, shift-select the control, and **Constrain > Parent**). Then freeze transforms, delete history and name it ***left_handControl***. Lock and hide all attributes.

Modify > Add Attribute…

Name: indexCurl

Make attribute: Keyable

Data Type: Float

Min Value: 0

Max Value: 10

Default Value: 0

Also add, with the same settings:

middleCurl

pinkyCurl

thumbCurl

Then set Min, Max and Default values to **-10, 10, 0** for:

thumbRotX

thumbRotY

thumbRotZ

fingerSpread

Now we'll use Set Driven Keys (SDK) to control the fingers. These are animation keys on attributes driven by the values of other attributes.

(**Animation** menu set) **Key > Set Driven Key>Set...**

Select *handControl* and click **Load Driver**. Select both the index finger joints (1 and 2) and click **Load Driven**. Highlight **Index Curl** under **Driver** and **Rotate Z** under **Driven** (with both joints highlighted). Click on **Key** to set the initial keyframe.

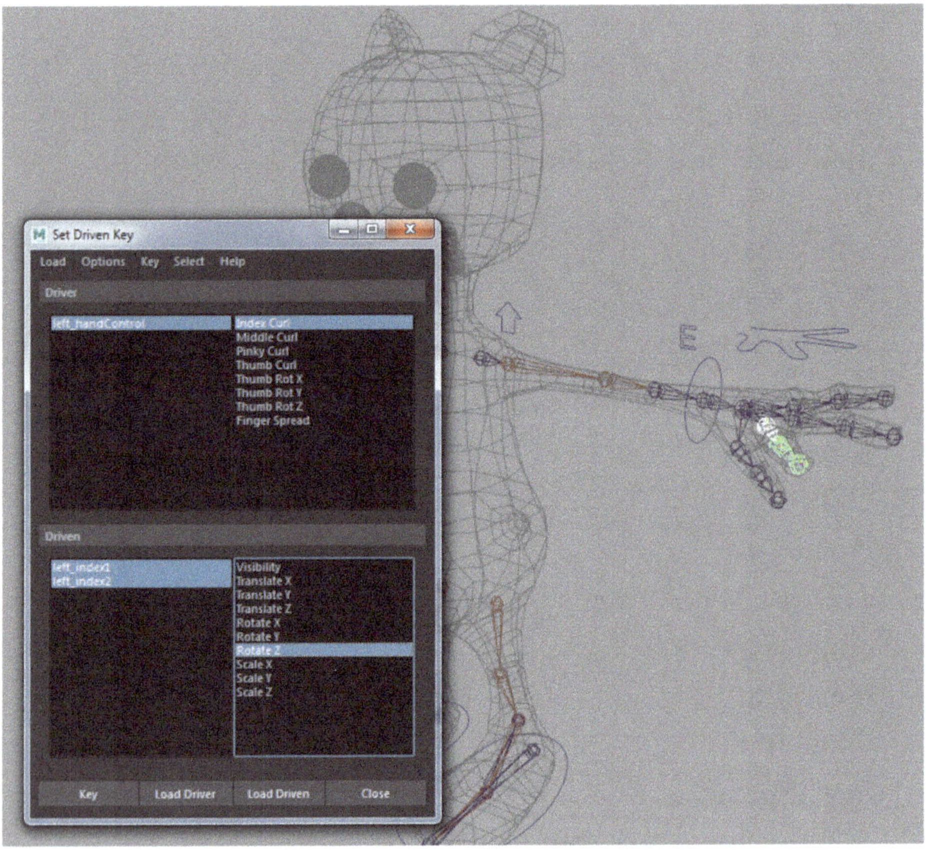

In the SDK window, select *left_handControl* and in the Channel Box set *indexCurl* to **10**. Click to highlight both joints under **Driven** and rotate them in Z in the viewport until the tip of the finger touches the palm. Press **Key**.

Test the driven key by selecting the *handControl* and in the Channel Box, highlight the **Index Curl** attribute and MMB-drag in the viewport to invoke the virtual slider. After testing, return the value to 0.

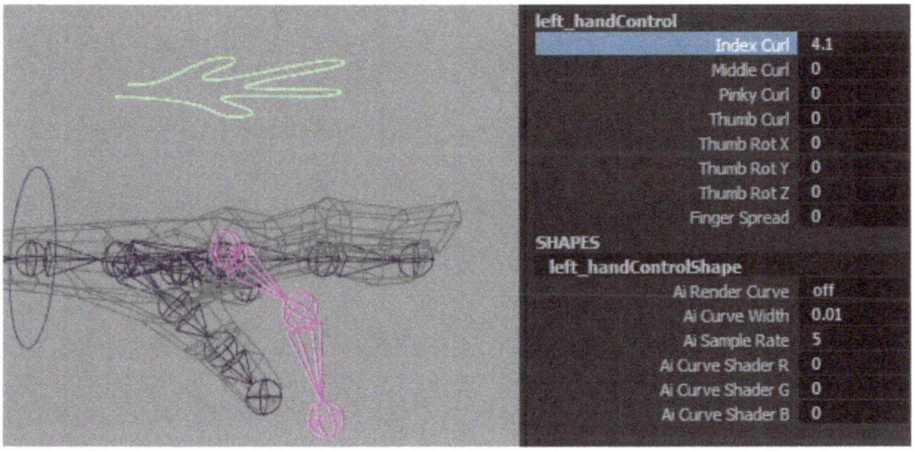

TIP

You can see the animation curve in **Window > Animation Editors > Graph Editor** with one of the driven joints selected. Sliding one of these keys vertically (hold Shift with MMB) will change the value.

Repeat this process for the other two fingers' curl.

Next, set *fingerSpread* = **10** and keep the three fingers far apart, keying *pinky1* and *index1* **rotateY.** Then move them close together for *fingerSpread* = **-10**.

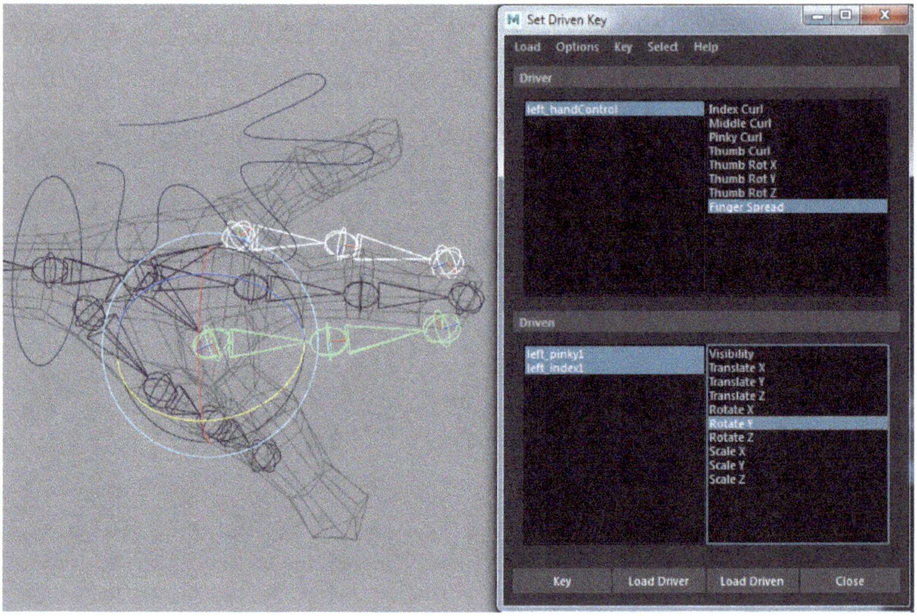

Before setting SDK on the thumb, go into component mode with the thumb joints selected and look at Local Rotation Axes (RMB over the question mark icon in the status bar's component selection mask area and choose **Local Rotation Axes** to view and manipulate them). You can use the rotate tool to line them up so that rotating in the Z axis will curl toward the palm (Y points up through the knuckles).

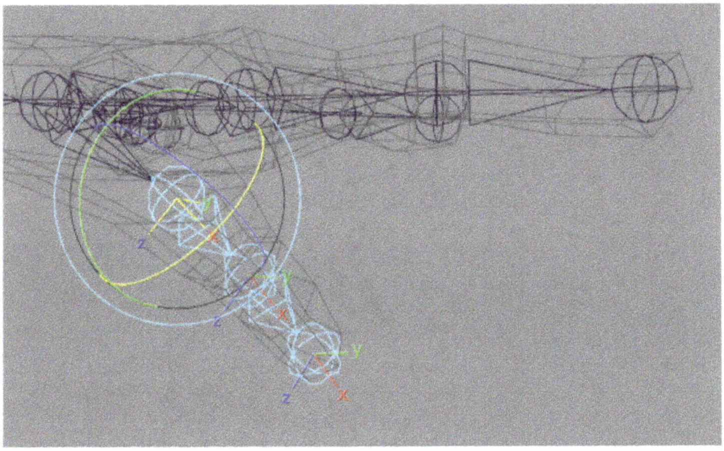

The thumb also needs special consideration, as it has more degrees of freedom than the other fingers. For this reason, you may find that, even on a toon-style character like this, a fourth joint is necessary to achieve the full range of hand shapes. We can add a joint to an existing chain by parenting. First place a joint near the wrist in the front view and enter to complete the operation.

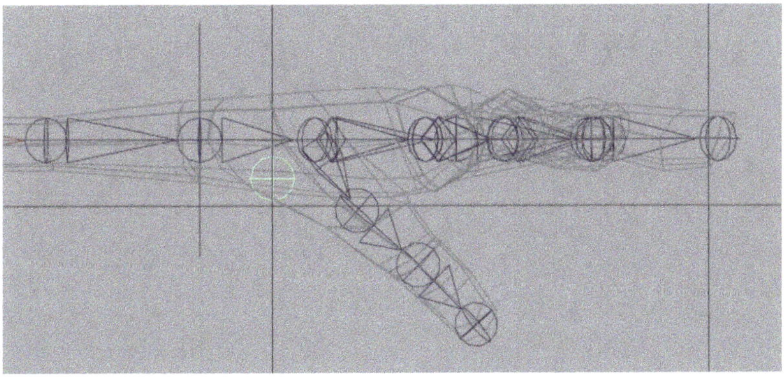

Use the top and side views to place this new joint near the lateral side of the wrist. We want the thumb to be able to rotate from this position in order to "cup" the hand.

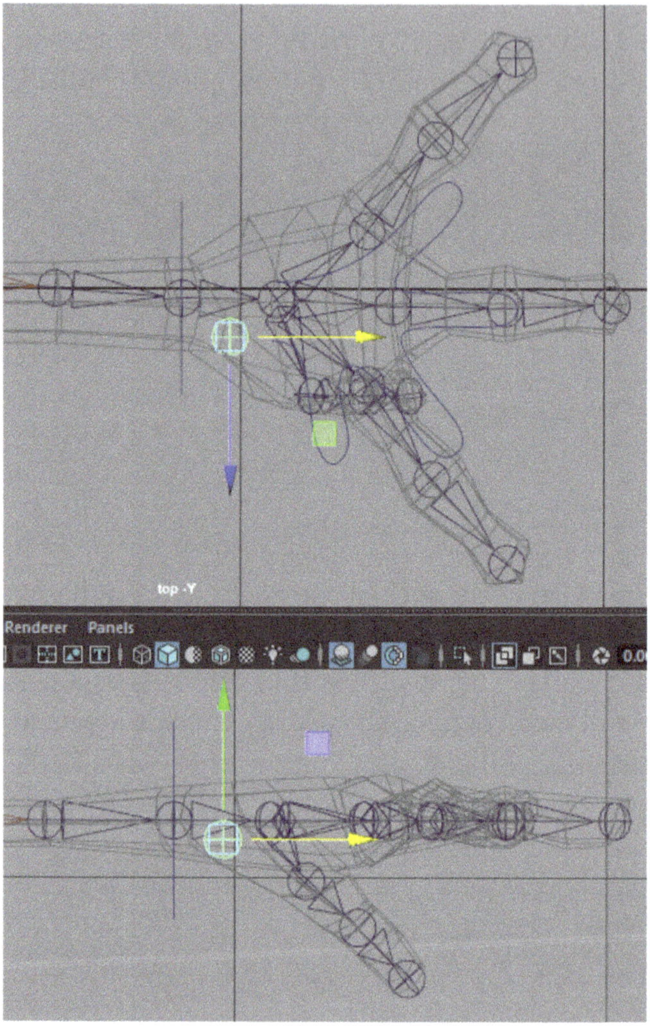

With this new joint selected, Shift-select the hand joint and hit **P**. This will make it the child of the hand joint. Next, select the first thumb joint and Shift-select the new joint and hit **P**. This will make the thumb chain parented under the new joint. You should end up with something like this:

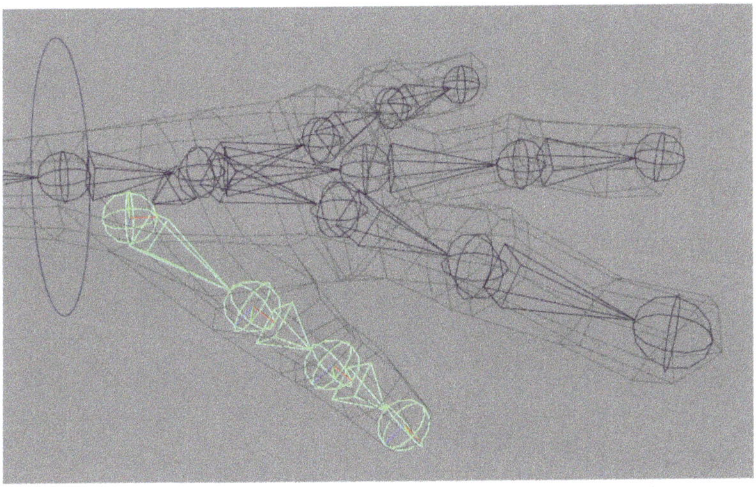

Rename the new joint *left_thumb1* and rename the other thumb joints *left_thumb2*, *left_thumb3*, *left_thumb4*. Rotate the local rotation axis to be oriented with the others.

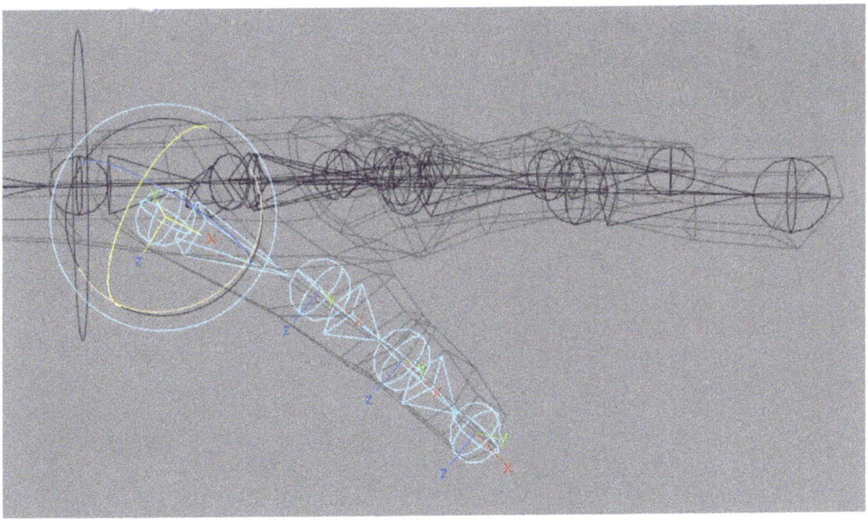

TIP

In some animation pipelines you may have to keep the **Rotate Axis** (which is what we're adjusting by rotating the Local Rotation Axes) of all joints "clean" (i.e., zeroed out). In those situations, you'd orient joints by typing directly into the **Joint Orient** fields in each joint's Attribute Editor. It is more time-consuming, but sometimes required, as Rotate Axis values don't always export correctly to game engines or other applications.

For the *thumbCurl* attribute, set **rotateZ** keys on the two middle thumb joints (*thumb2*, *thumb3*).

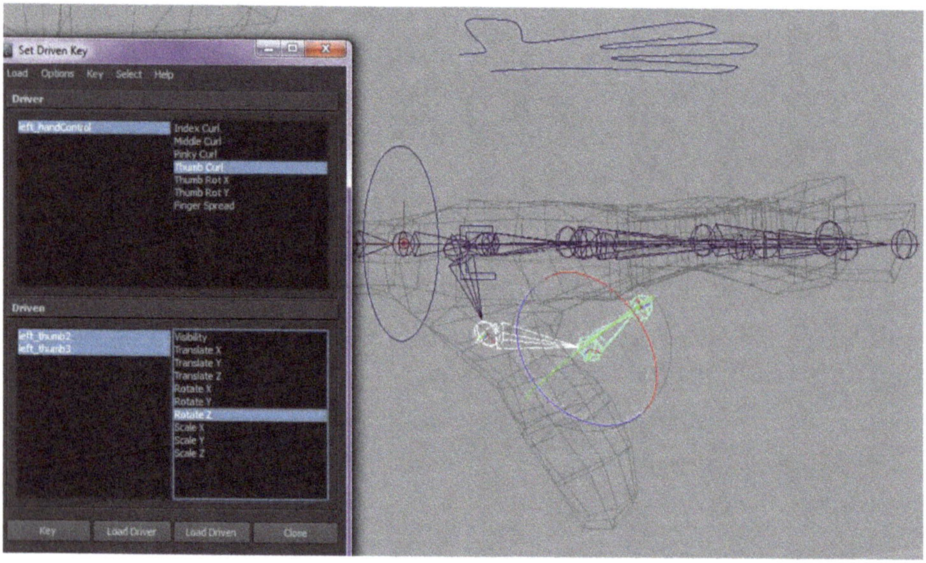

Set SDK on only the first (*thumb1*) joint for **Thumb Rot X**, **Thumb Rot Y**, **Thumb Rot Z**, to control this joint's rotation in X, Y, and Z respectively. These extra controls are necessary to get precise thumb angles for, say, holding onto objects. Set keys at **0** (default), **-10,** and **10** values for each attribute.

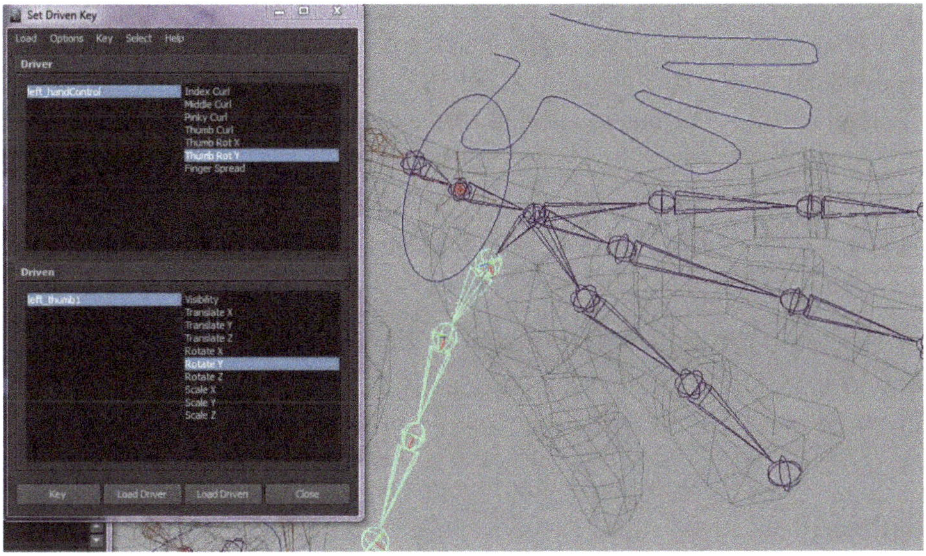

Select the top of the joint hierarchy (*clavicle*) and **Mirror Joints**, remembering to set the options to search and replace prefixes. You also need to create ***right_armIK*** since IK only mirrors if it's attached to the root joint you're mirroring (which ours is not). Don't forget to move the end effector.

TIP

The **Mirror Joints** function in Maya sometimes copies over some junk. In the Outliner, Shift-click the plus sign next to the top joint in the mirrored hierarchy (*right_clavicle*) to reveal the entire hierarchy and delete anything that's not a joint; i.e., unused constraint nodes and end effectors.

Set up *armControls*, *clavicleControls*, *elbowControls*, and *handControls* for the right side as well (review sections 3.3 and 3.4).

TIP

To flip copies of your controllers over to the right side, select all arm control groups together, group them, then **Duplicate Special** with scaleX -1. The elbow "E" control might need re-flipped to not appear backwards. If you use this method, you'll get errors when you try to freeze transformations on controls with locked transform attributes. Use the rt-click menu over attribute names in the channel box to selectively freeze transforms. If you copy your controls, they'll already be in the right places with the right orientation and attributes; you'll just have to create all the connections.

3.5 CREATING THE SPINE AND HEAD JOINTS

To setup the spine we'll introduce deformers for the first time, so now might be your last chance to easily **Edit > Delete All by Type > History** and **File > Optimize Scene Size.**

In the side view, start by placing a root joint just above the hip joints of the legs.

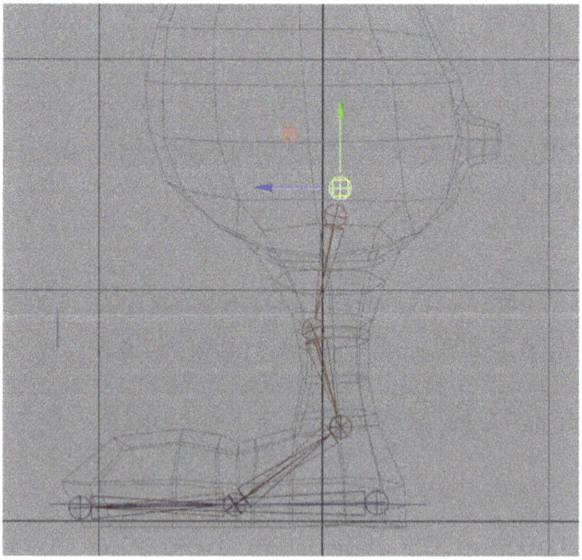

From there, draw an s-curve of evenly spaced spine joints up the back, as in the picture below. 6 or 7 joints should do.

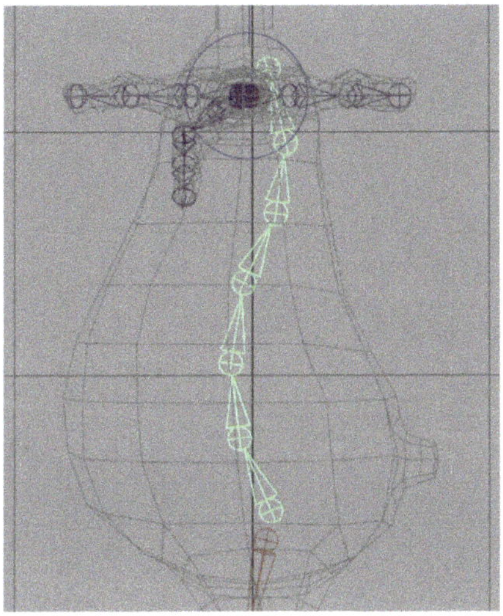

Name the first joint ***root***, then ***spine1, spine2, spine3***, etc.

Use the joint tool to click directly on the last spine bone to continue the chain up the neck. Two or three neck joints should do. Name them ***neck1, neck2***, etc.

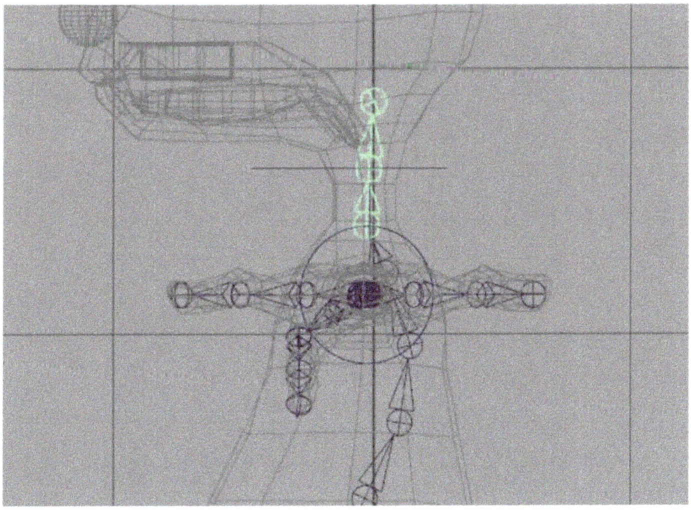

Then do ***head***—right in the center of the head.

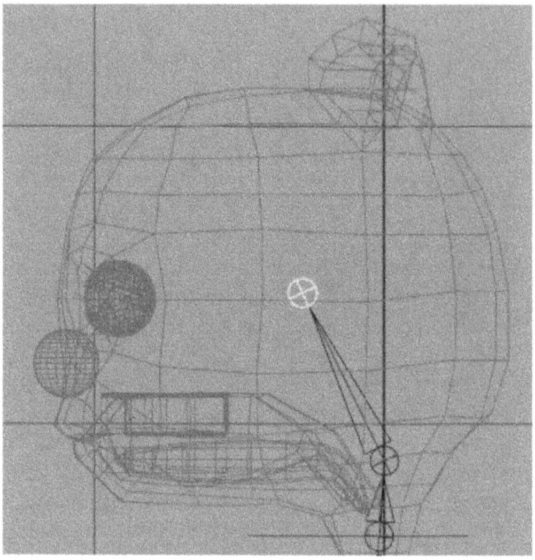

Next do ***jaw*** and ***chin***. Jaw should be placed on a realistic human character right where the mandible hinges to the skull, just in front of the ear—the origination of the jaw rotation. For our cartoon character, somewhere behind the corner of the mouth will work. These two joints allow us to rig the jaw opening and closing.

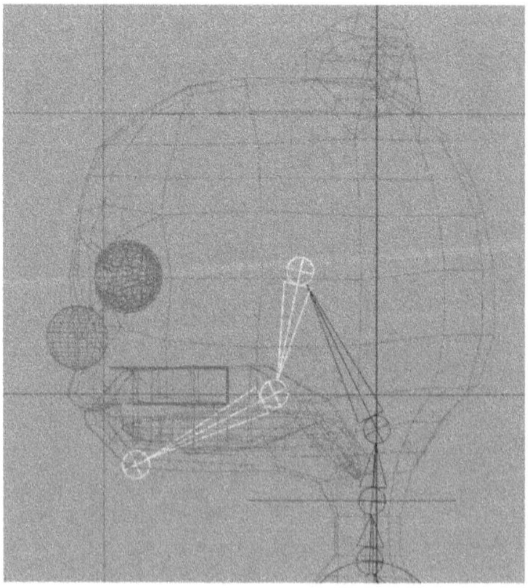

Now we'll create joints to drive the eye rotation. These have to be in the exact center of the eye geometry. Start by creating a joint anywhere and then zeroing out all its translation values in the channel box. You should find it sitting at the world origin. Untemplate the character's eye geometry so you can select it. Select the joint, then the eye geometry, and invoke **Modify > Snap Align Objects > Align Objects (Options)** and set it to **Mid, World X, Y, Z** and **Last Selected object**. The joint should pop to the center of the eyeball. (Align Objects only works with joints if they are zeroed out first.)

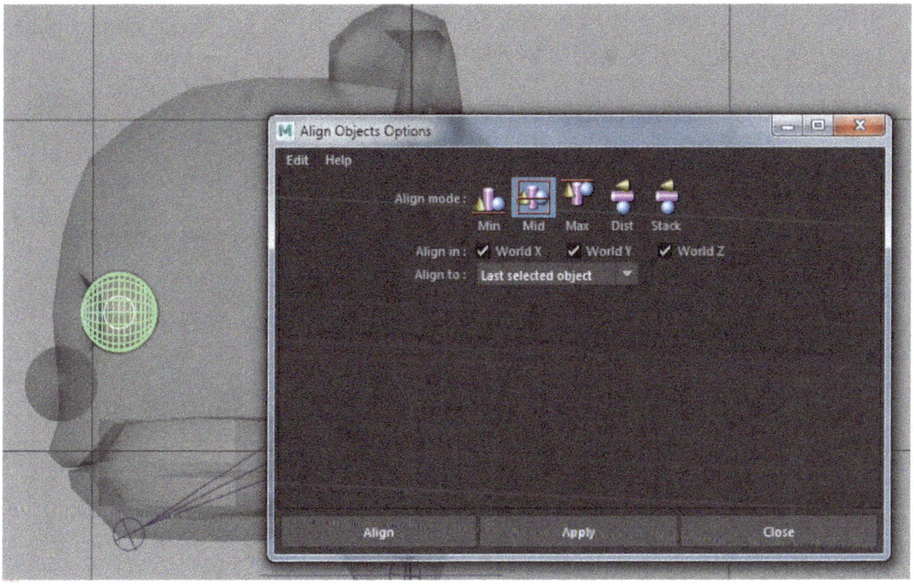

Do this for the other eye. Name these joints ***left_eye*** and ***right_eye*** and parent them to ***head***.

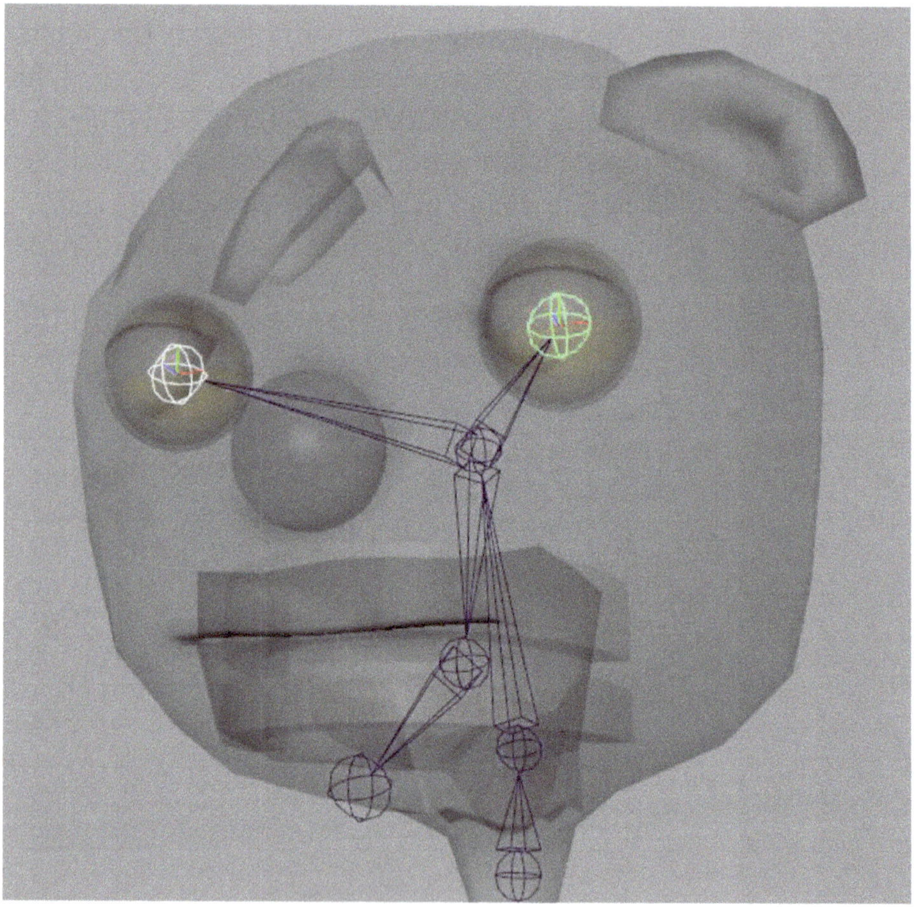

3.6 SPINE SETUP

Now check the local rotation axes in the spine. With the spine selected, switch to component mode and right-click the question mark on the status line to be in local rotation axis mode. Ideally, all of the secondary axes (Y) should be facing the same way—we see that half of them are flipped.

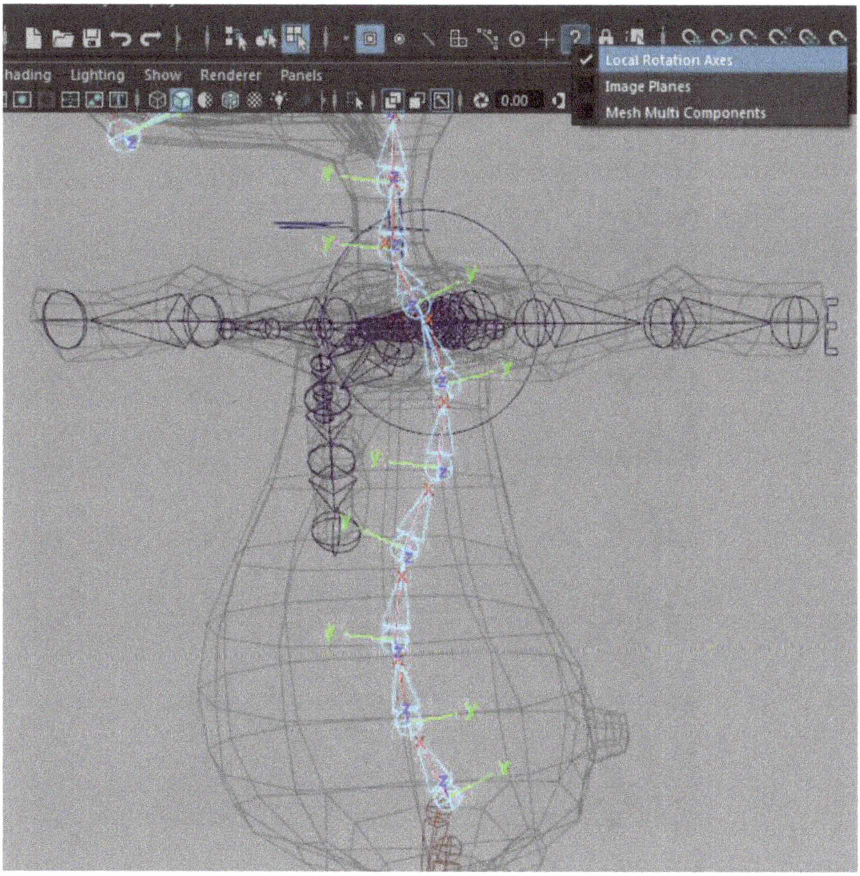

Select any spine joint rotation axes that are facing the wrong way and type the following in the MEL command line:

rotate -r -os 180 0 0;

This command will rotate 180 degrees in X in relative (-r) object space (-os) mode. You can MMB-drag this command from the script editor to the shelf to make a handy button to execute this. Line up all the Y axes in the same direction. Then leave component mode.

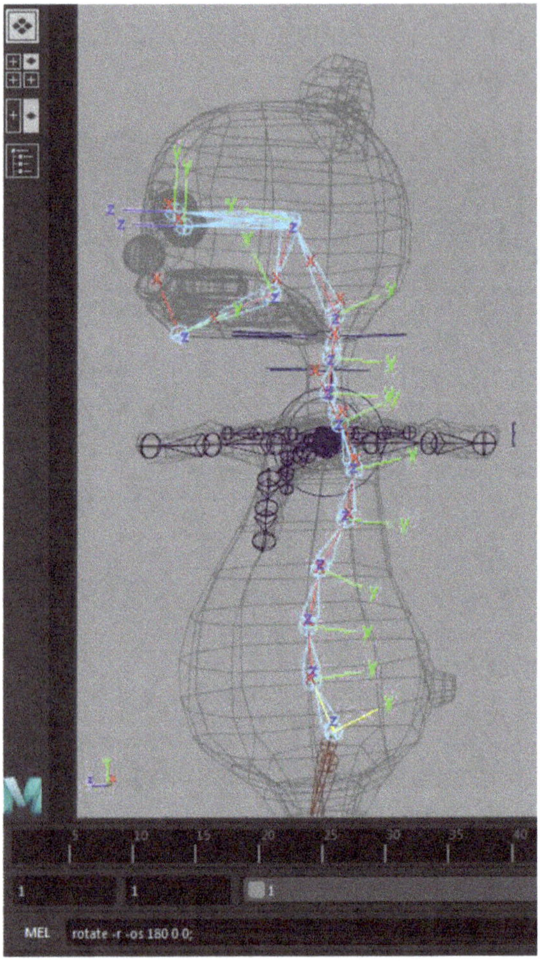

Skeleton > Create IK Spline Handle (options). Turn off **Auto parent curve**. Click on first spine joint (the one above the root) and then click on the last (or second to last) spine joint to set an IK spline handle. Rename it **spineIK**.

TIP

We will be creating clusters out of the curve control vertices which we will parent to animation control objects, so we didn't want to **Auto parent curve** or this would result in double transformations when the character rig is translated. We could have also kept ON **Auto parent curve** and later turned on **Relative** in the cluster attributes to compensate.

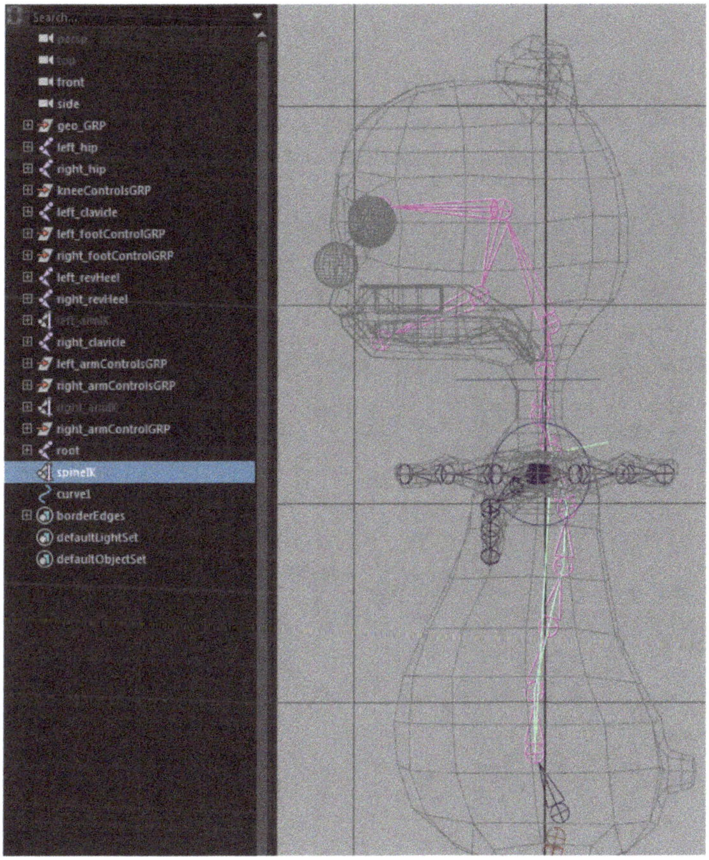

This type of IK creates a NURBS curve running down the center of the joint chain which you can manipulate to shape the joint chain. We'll **Cluster** the CVs to have a way to manipulate the curve CVs in object space.

Select the back curve—find it at the bottom of the outliner called *curve1*—rename it ***backCurve*** and go into CV mode. The curve should have four CVs: one at the bottom, one at the top, and two in the middle. Select the bottom two CVs (they look like a square and a U) and **Deform > Cluster** (make sure **Relative** is **OFF**). Do this for the top two CVs as well. Rotate and translate the "c" icons to test your IK back rig.

TIP

It's easiest to select clusters in the Outliner, as they are low on Maya's default selection priority (selecting a cluster overlapping a joint will default to selecting the joint).

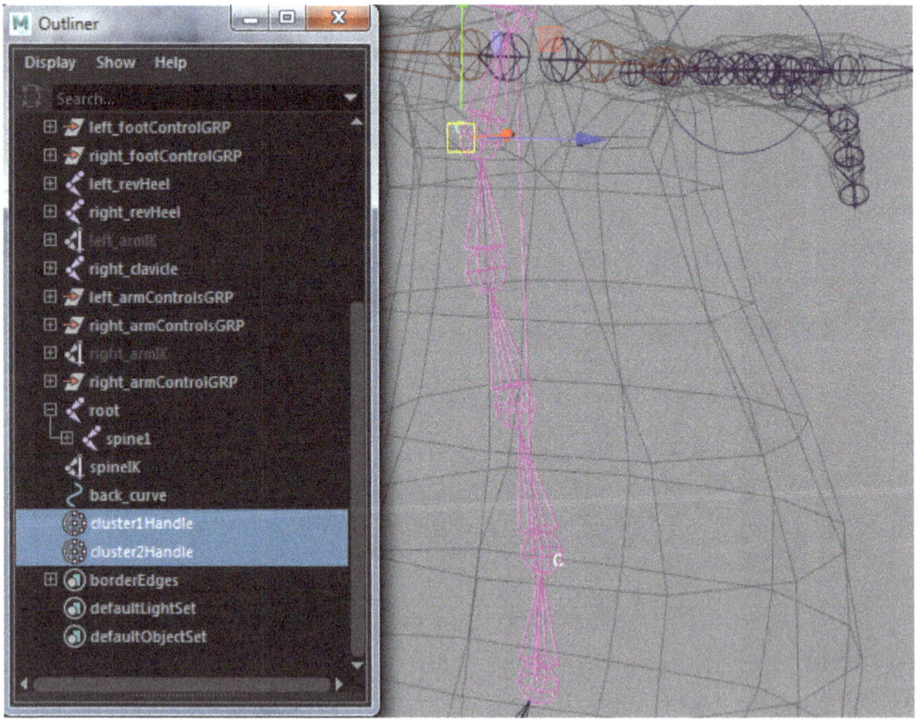

The bottom cluster (*cluster1Handle*) won't be animated, so parent it (with **P**) to the *root* joint (this automatically creates a group node above the cluster because the cluster isn't ***Relative***—the group preserves its world-space position). You can also turn off its **Visibility** and lock and hide all its attributes.

Next make another IK Spline Handle from the base to the top of the neck. Don't start on the same joint your spine IK ended on or you'll be sorry. Instead, start at the next joint up. Create two clusters, one from the top three CVs and one from the bottom CV. Again, parent the lower cluster to the joint below the IK handle (spine 6 or 7) and hide it. Name all of your new nodes in the Outliner (***neckIK, neckCurve, neckUpperCluster***, etc).

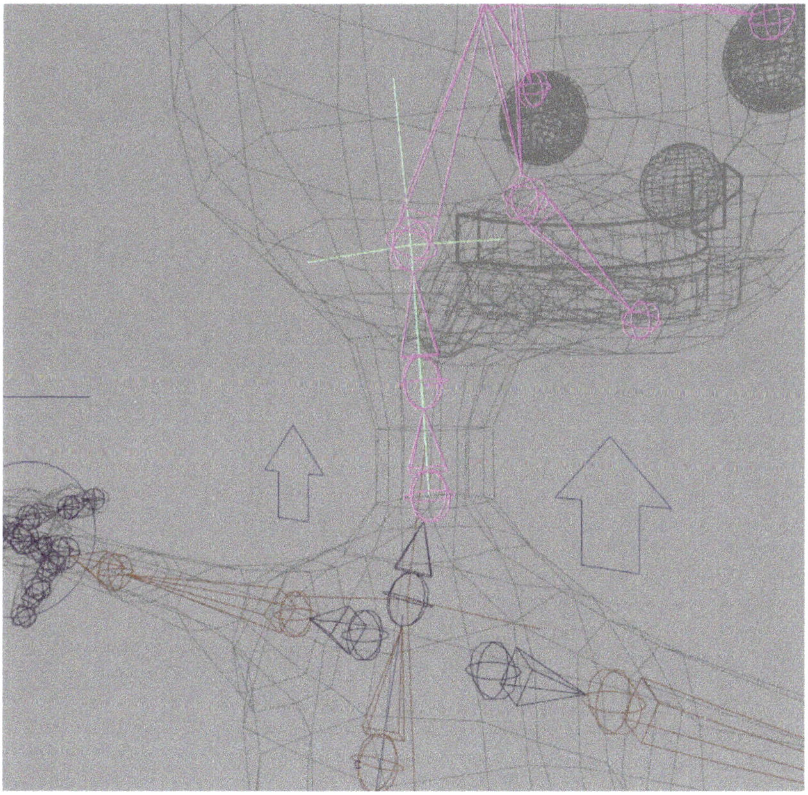

Create a polygon cube and move it to an empty area. Then go to **Create > Curve Tools > EP Curve Tool (options).** Set it to **1 Linear.** Hold **V** and snap a curve to every vertex of the cube, making sure to get a curve along every edge. Then delete the poly cube and you are left with a NURBS curve cube you can use as a control object.

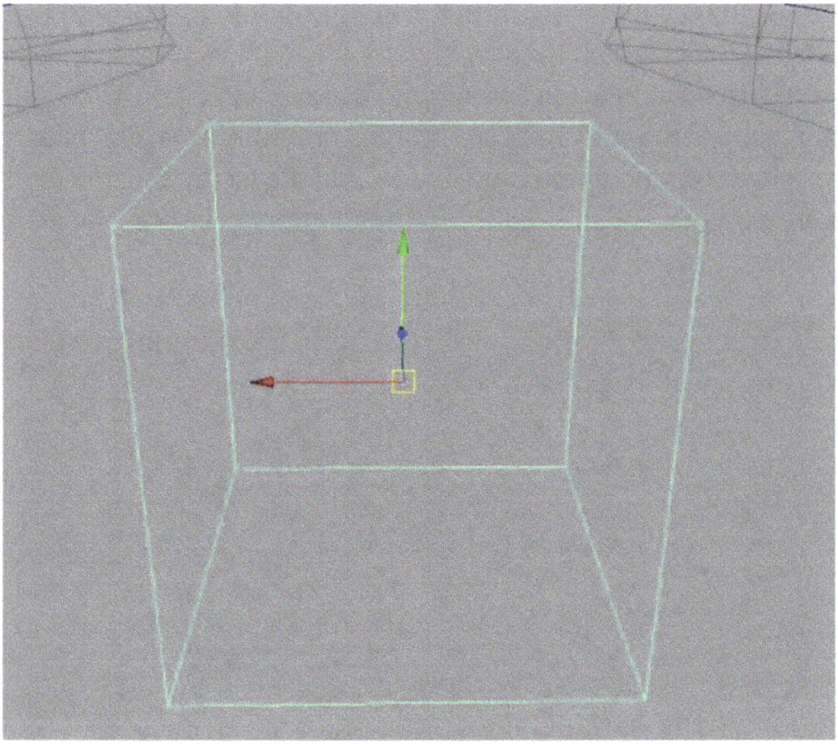

Move and scale this to surround the upper torso (**Modify > Center Pivot** and **V**-snap to the upper spine cluster, then scale and move CVs to shape it). Then Ctrl+D to duplicate and **V**-snap to the *root* joint, fitting it around the hips. Name the upper ***spineControl*** and the lower ***rootControl***.

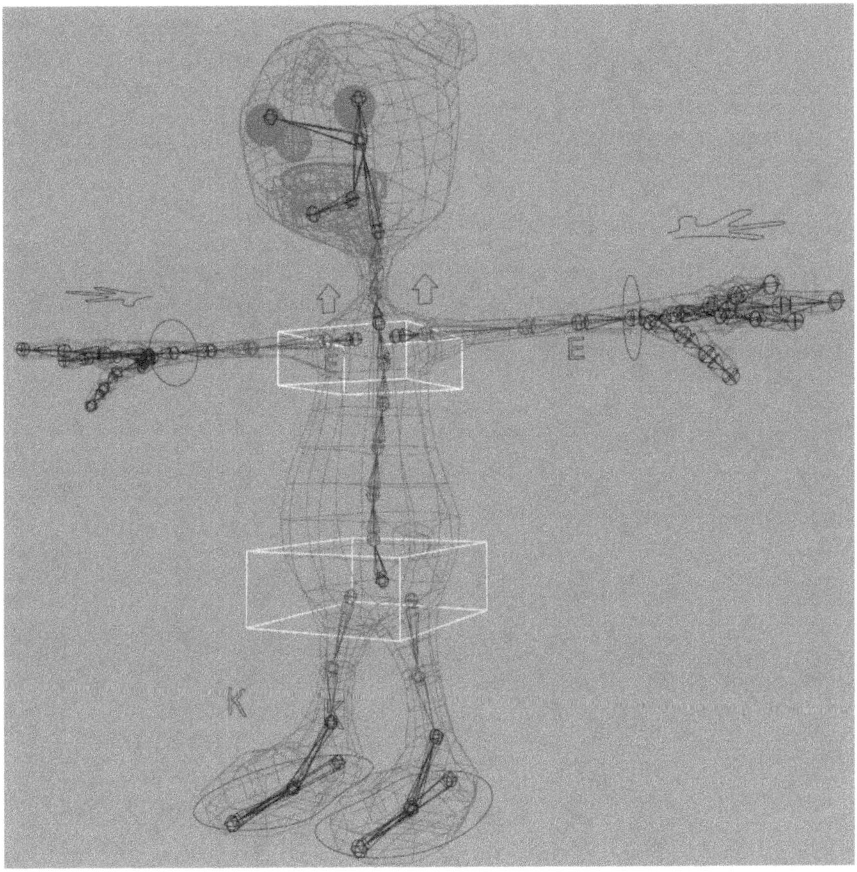

Now create two NURBS circles and position them around the neck and the waist at the root joint. Name them ***neckControl*** and ***hipsControl***. Freeze transformations and delete history on all new control objects.

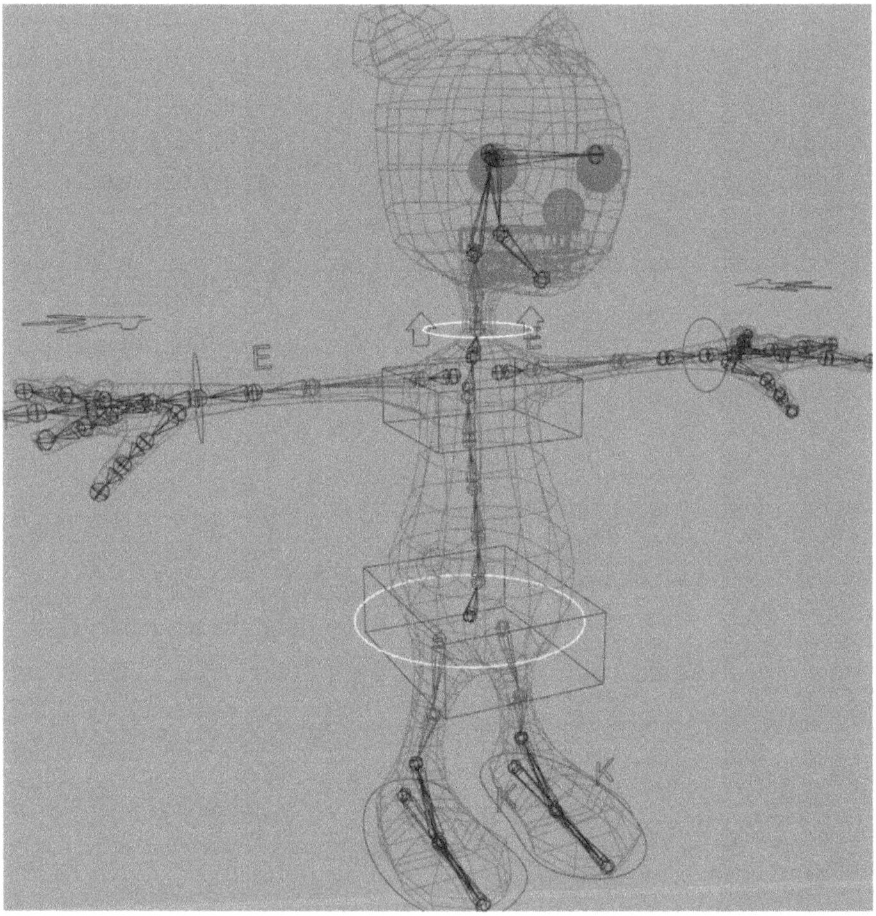

Now to rig 'em up. Select *spineControl,* and make sure its pivot is snapped to the spine cluster. Then select the cluster, Shift-select the *spineControl* and hit **P** to parent. Test by moving and rotating the *spineControl* to move the spine, then return to **0**. Lock and hide **Scale** and **Visibility** attributes.

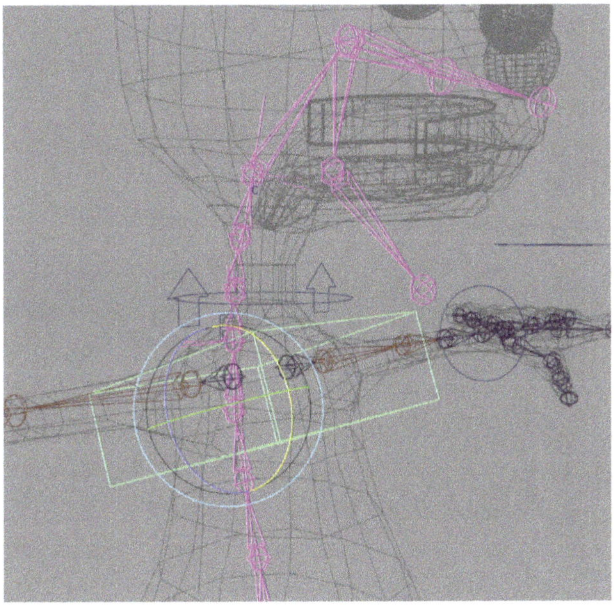

With *spineControl* still selected**, Modify > Add Attribute…** and add ***twist*** and ***ikFkBlend***. Set min, max, and default values of *ikFk-Blend* to **0**, **1**, **1**. Open Connection Editor and load *spineControl* on the left side, *spineIK* on the right. Connect ***twist*** with ***twist***, ***ikFkBlend*** with ***ikBlend***. Test these attributes by selecting the *spineControl* object and using the new **Twist** attribute to rotate the spine and **Ik Fk Blend** to turn off the IK for direct manipulation of the spine bones.

Follow the same procedure for the neck: Snap *neckControl* pivot to the *neckCluster* (hold **D** and **V**), parent the cluster to the *neckControl*, lock and hide **Scale** and **Visibility**, add attributes for ***twist*** and ***ikFkBlend*** and connect them to the *neckIK*'s ***twist*** and ***ikBlend***.

In the side view, select the Create Joints tool and click on the root joint to highlight it. Then place a single joint just below it, centered between the hip joints. Call this ***pelvis***.

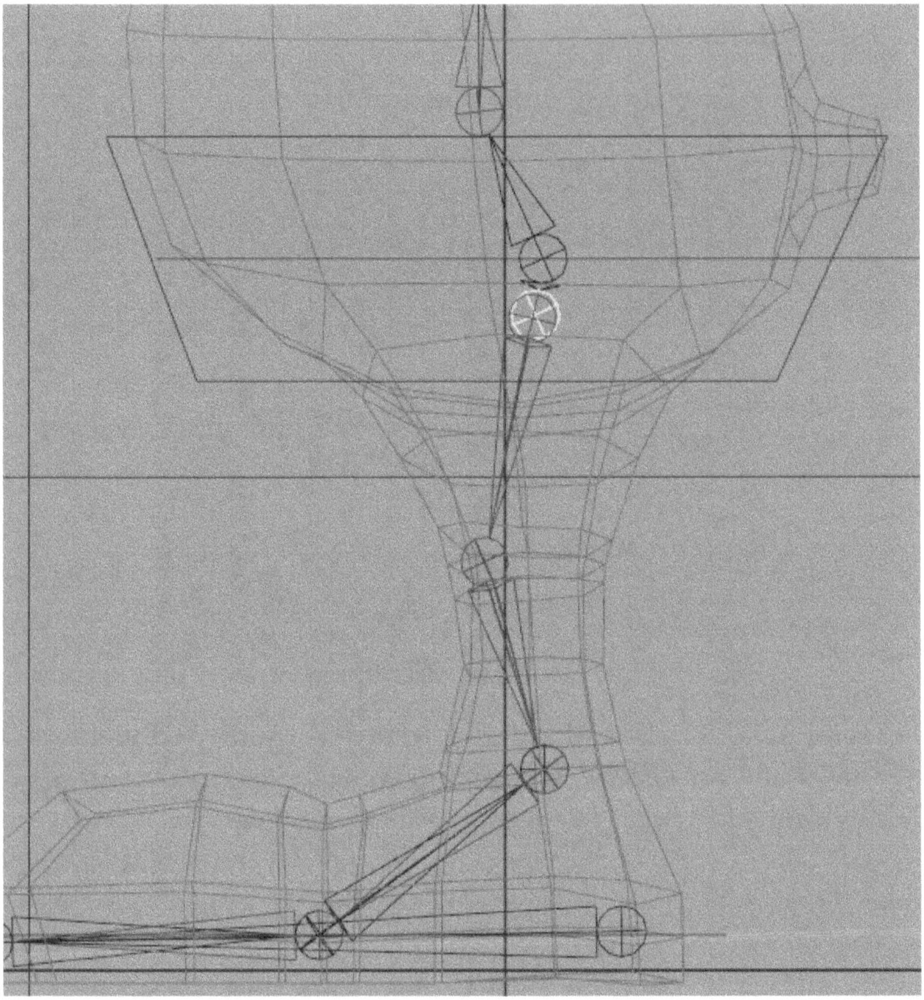

Select both hip joints and then the pelvis and hit **P** to parent.

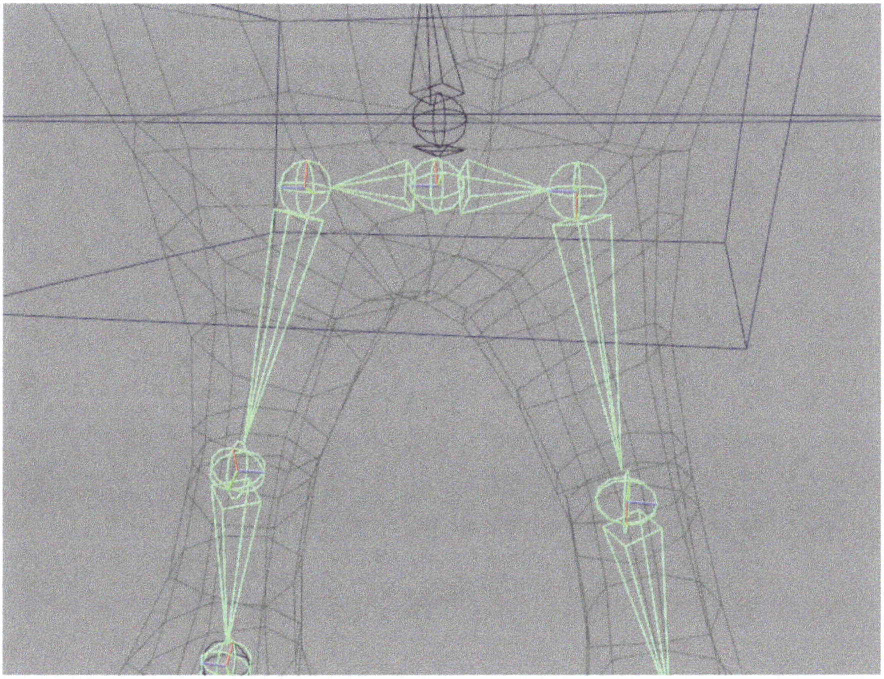

Select *hipsControl* and snap its pivot to the pelvis joint. Select the control and then the *pelvis* joint and **Constrain** > **Orient**. Lock and hide **Translate**, **Scale**, **Visibility**.

Select *rootControl* and snap its pivot to the root joint. Select the control and then the *root* joint and **Constrain** > **Parent**. Lock and hide **Scale** and **Visibility**.

3.7 BRINGING IT ALL TOGETHER

From the top view, draw a four point arrow like this:

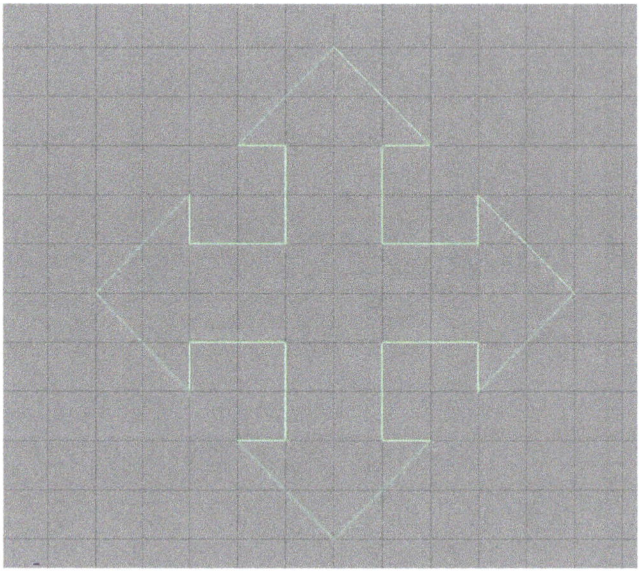

Center pivot, scale down, and snap it to the world origin:

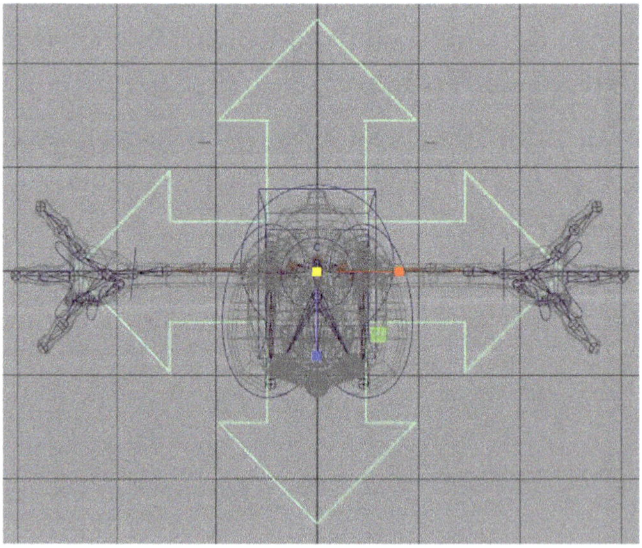

Freeze its transformations. Rename it to ***masterControl***. Lock and hide **Scale** and **Visibility**.

In the outliner, Ctrl-select *left_footControlGRP*, *right_footControlGRP* (the groups above the controls), *rootControl* and finally *masterControl* and hit **P** to parent them.

Select *hipsControl* and parent to *rootControl*.

Select *right_kneeControl* and *left_kneeControl* and parent to *rootControl*.

Select *spineControl* and parent it to *rootControl*.

Select the *left_clavicle* joint and *right_clavicle* joint and parent them to the joint above the *spineIKCluster* (*spine6*?)—the joint the *spineIK* handle is attached to. If the clavicle joints appear to "flip," it is because they are trying to auto-orient to their new parent joint and they can't because they have an incoming connection—the parent constraint to the *clavicleControl*. If this happens, find the parent constraint in the Outliner (below the *clavicle* joint), select it and hit the **Delete** key, parent the joints again (they won't flip this time), and then reconnect the *clavicle* joints to the *clavicleControls* with new parent constraints.

Select the groups above *left_clavicleControl* and *right_clavicleControl*, *neckControl* and then parent them to *spineControl*.

Select *spineControl*, then the joint the *spineIK* handle is attached to and **Constrain > Orient**.

Select the groups above *left_armControl*, *left_elbowControl*, *right_armControl* and *right_elbowControl* and parent to *masterControl*.

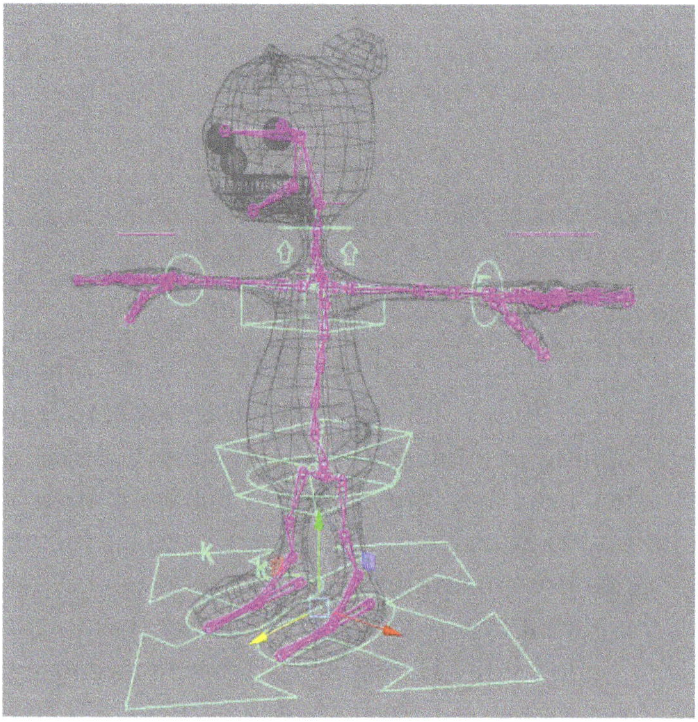

You should be able to move and rotate the *masterControl* now with the whole skeleton following.

3.8 HEAD AND EYES SETUP

Create a NURBS object and snap its pivot to the joint that starts at the base of the head (in my case ***neck3***). I made a little pyramid and placed it over his head:

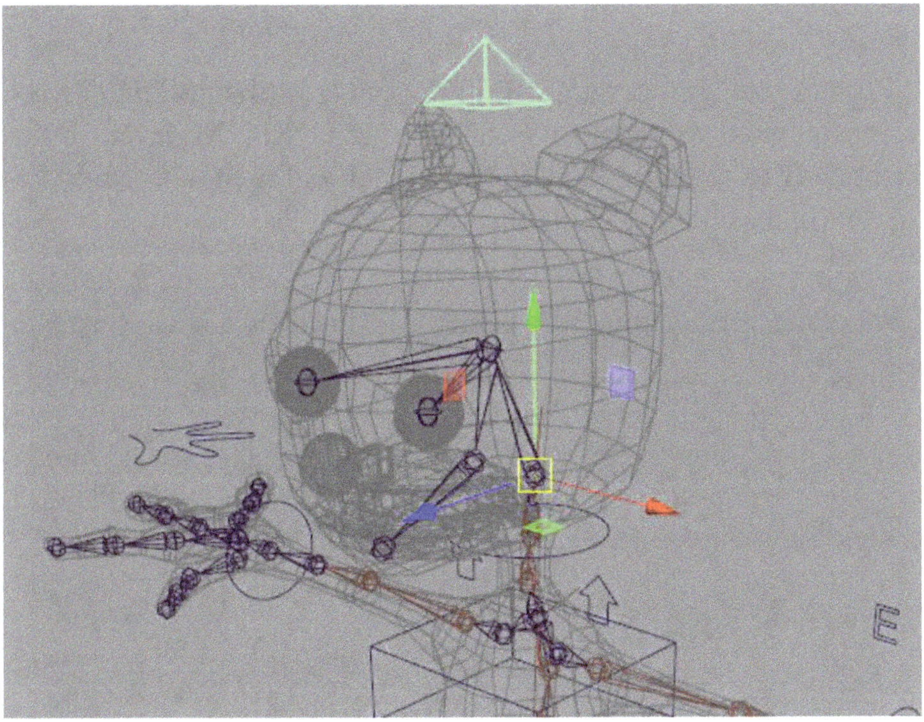

Name it ***headControl***. Freeze transforms and then orient constrain the same joint (***neck3***) to the *headControl* (remember to **Maintain offset** in the options). Parent the *headControl* to the *neckControl*. Lock and hide **Translate**, **Scale**, and **Visibility**.

TIP

You can optionally add a Boolean *headOrient* attribute to the *headControl*, connected to the orient constraint's **Head Control WO** attribute, to be able to turn on or off orientation driven by the head control. By turning it off you'd basically be using the neck control to orient the head.

Check that the eyes are still in the center of the *eyeGEO*. If not, you can create a temporary Point Constraint with **Maintain offset** turned **off** to snap the joints to the center and then delete the constraint (in the Outliner under the joint).

Now make another NURBS circle, transform it into position in front of the eyes and pinch its center UVs to create a mask shape:

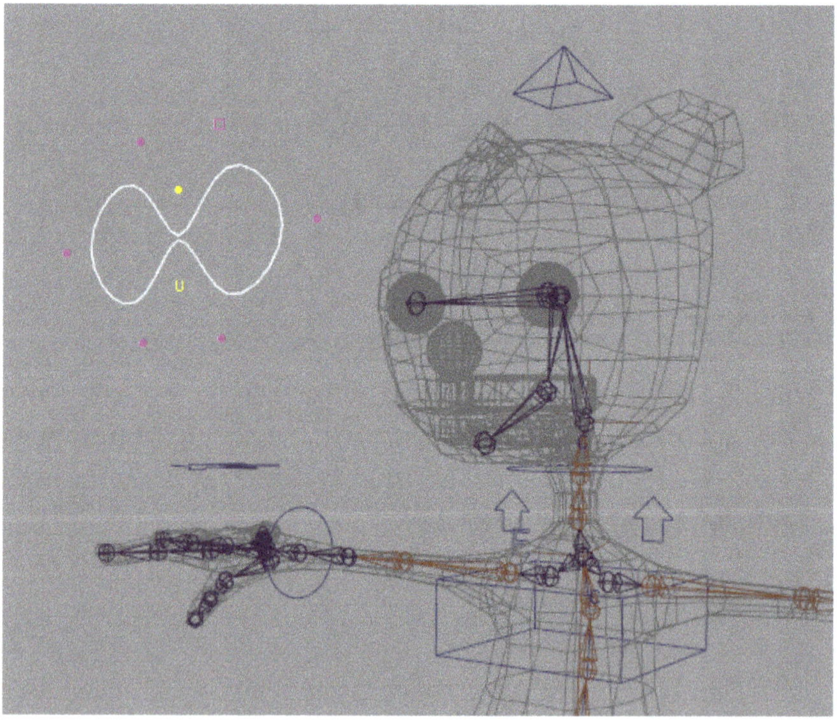

Name it ***eyesControl***. **Create > Locator** and snap it to the left eye. Drag it out front in Z, and holding the **C** key with the Z manipulator handle still highlighted, MMB-click on the *eyesControl* to snap it to the same Z-plane as *eyesControl*. Name it ***left_eyeControl***. Duplicate it and snap the duplicate to the right eye, moving it into the *eyesControl* Z-plane as well, and naming it ***right_eyeControl***. Resize the *eyesControl* to fit around the two locators, as pictured below. Freeze transformations and delete history on all three, and then parent the locators to the *eyesControl*.

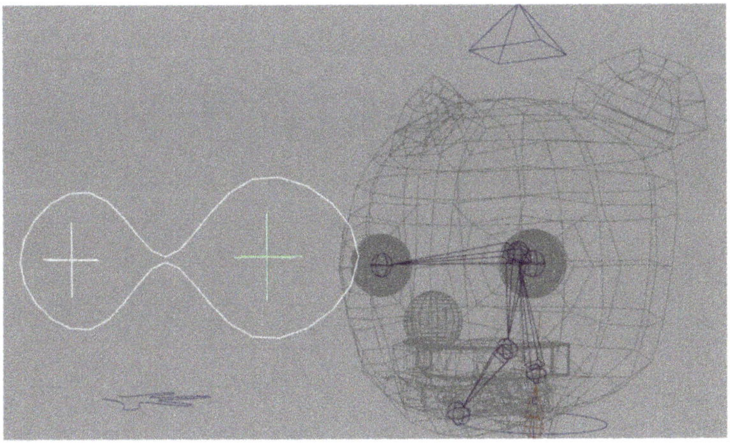

TIP

To control the size of a locator, adjust the **Local Scale** attribute under its Shape node.

Select the *left_eyeControl* and then the *left_eye* joint. **Constrain > Aim (options)** and set to **Maintain offset**. Do the same for the right eye.

Parent the *eyesControl* under the *masterControl*. (Some animators prefer to parent the *eyesControl* under the *headControl*, but I find this encourages an overall "spaced out" look in the CG character—the eyes shouldn't be following the body movements, they should typically be locked in focus on something. In animation, you can use this totally independent *eyesControl* to snap or constrain it to objects in your scene so the character appears focused and alert.)

To test these eyes out, go ahead and skin them to the eye geometry. Select the left eye joint and the left eye geometry, and go to **Skin > Bind Skin (options)**. Set **Bind to: Selected joints.** Do the same for the other eye and then move *eyesControl* to see the eyes rotate.

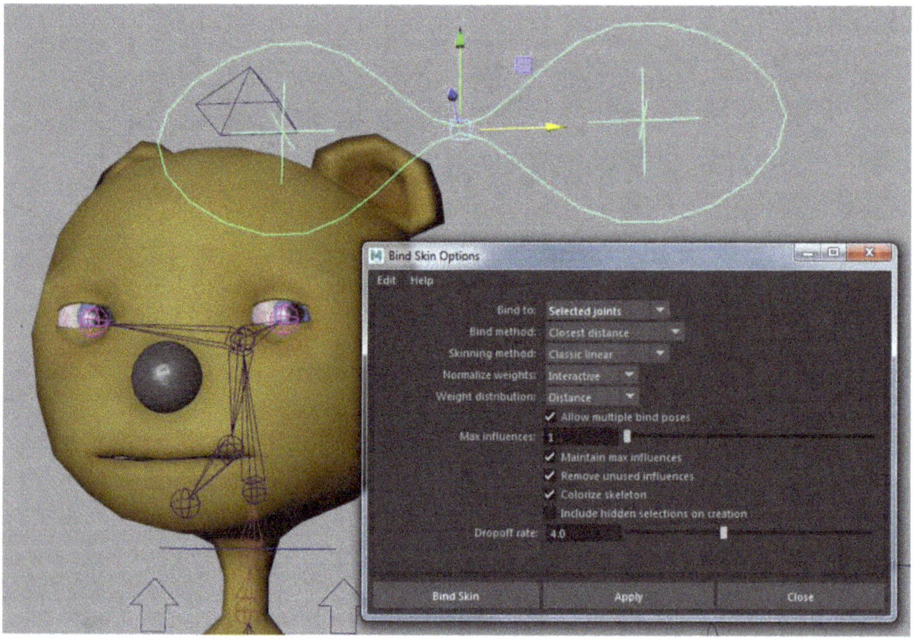

Lock and hide **Rotate**, **Scale**, and **Visibility** for all eye controls.

3.9 CLEAN UP

Now go to your Outliner and make sure **Display > DAG Objects Only** is on. Delete any empty groups and unused nodes. Select *masterControl* and group it—name this ***controlGRP***. Select the root joint of the skeleton and group it—name this ***skelGRP***. You can move the reverse feet into this new group. Select all the geometry and group them—name it ***geoGRP***. Select any IK splines or any other nodes that are left over and group them—name it ***miscGRP***. You should have four groups at the top of your scene graph now.

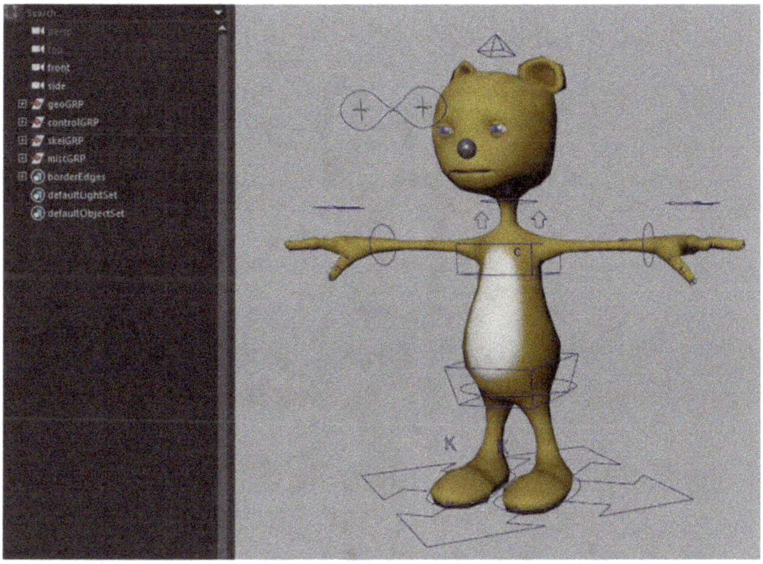

You could group these four into a main "generikat" group so that it remains discrete when you import or reference this rig into other scenes, but you could also have this happen automatically upon import. You may also be utilizing Maya Assets in your pipeline which will require a little further setup. For best practices, try to keep all of your control objects in one hierarchy, skeleton in another, and geometry in another.

Check once more that every control object is zeroed out and that you have locked and hidden any unused attributes. Save your scene as a backup, and then **File > Optimize Scene Size** and **Edit > Delete All by Type > Non-Deformer History**. Organize your layer editor so that all geometry is on a layer so you can later make it a non-selectable reference layer, and everything "hidden" (like IK handles, joints, etc.) is on its own layer that you can hide.

It's also a good idea to color-code your control objects, because pole vector objects and hand controls get all mixed up during animation. Select each of the left-side objects and color them blue by going under their attribute editors to **Drawing Overrides** and turning on **Enable Overrides**—then adjust the color slider to blue. Do red for the right side and green or yellow for the middle.

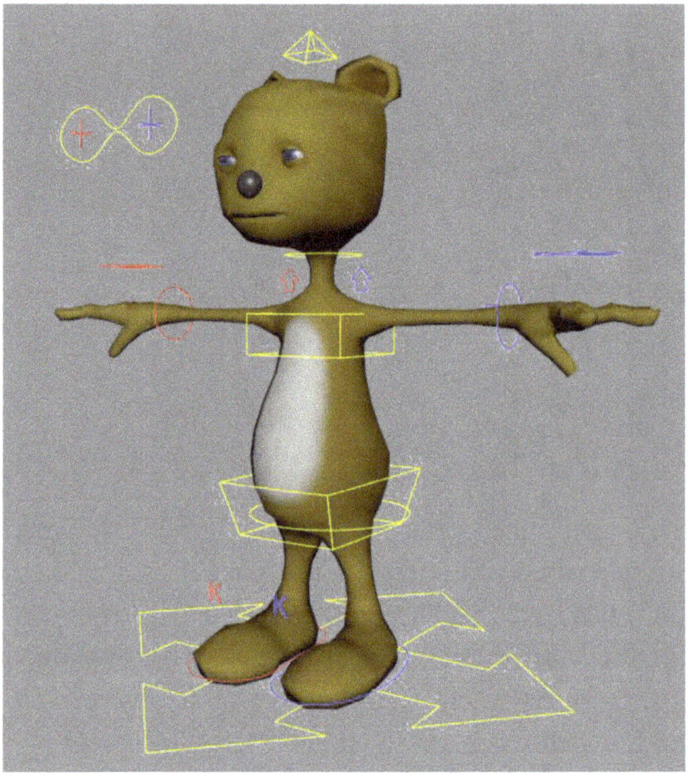

TIP

You can also assign all left-side controls to a display layer and assign that layer's **Color** to blue, which automatically makes their wireframes blue.

You are ready for skinning the mesh to the rig and facial setup, subjects we will visit in the next lessons.

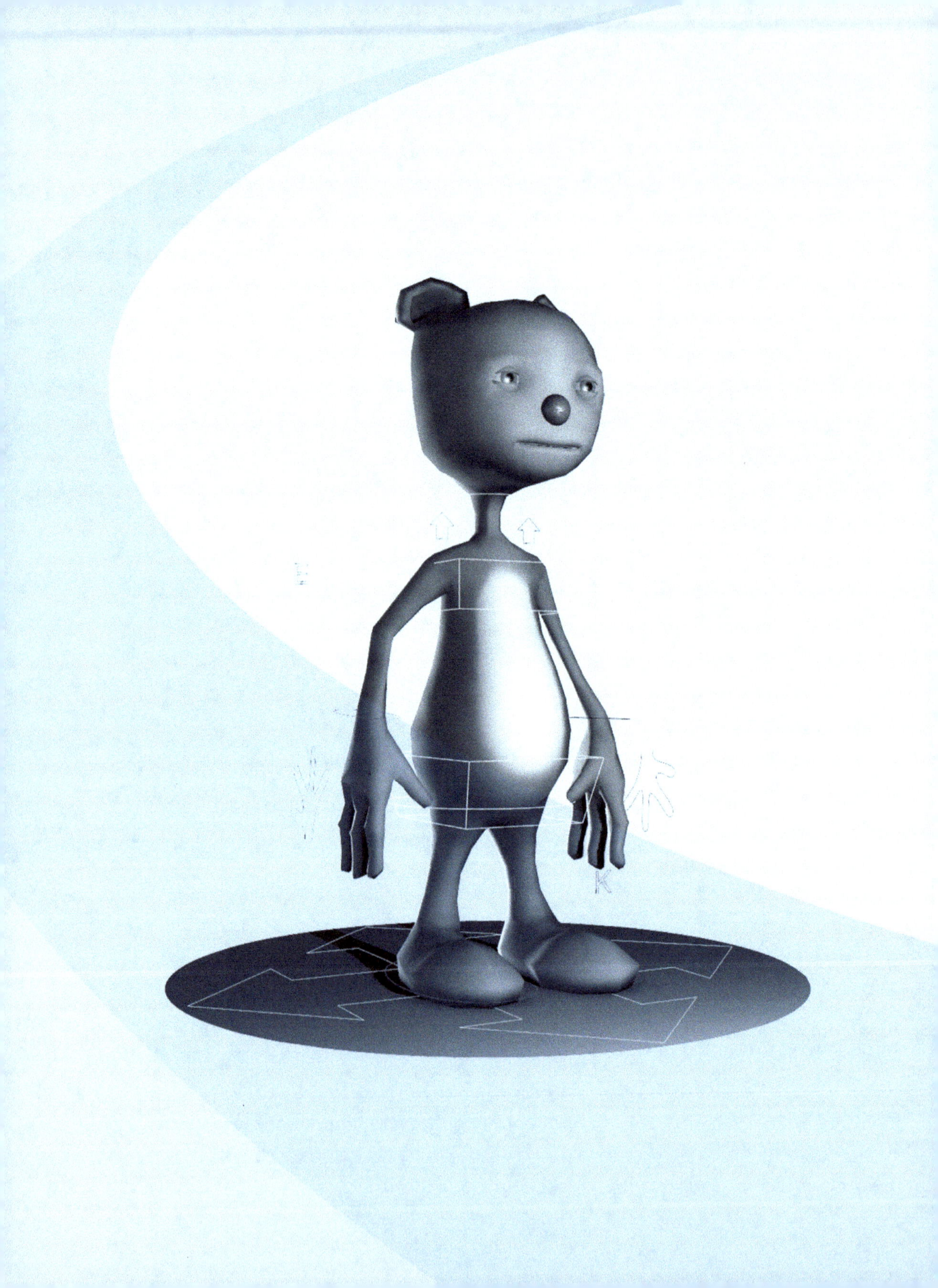

Part IV
Character Skinning

4.1 INTERACTIVE BIND SKIN

When we last met Generikat he was rig-ready and geometry-ready, but the geometry was not yet bound to the rig. Let's do that now.

Select the *root* joint and Shift-select the geometry and **Skin > Interactive Bind Skin (options)**. Set **Bind method** to **Closest in hierarchy** and **Max influences** to **3**. These settings work better for a very simple character such as ours.

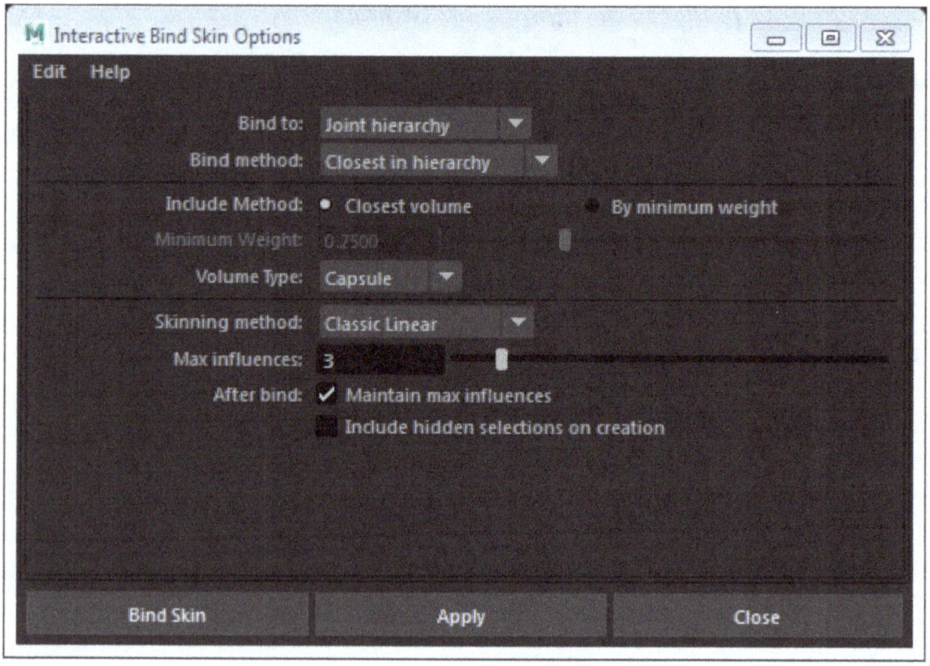

TIP

Use the panel menu **Shading > X-Ray Joints** mode to be able to see and select joint in shaded mode.

Now a capsule appears around the root bone of your character. You should see “root” highlighted under the **Influences** section of the **Tool Settings** window. The capsule gives you a visual representation of the influence envelop around each joint. Successively click through the influence list, selecting each bone and adjusting the capsules size so that they more tightly fit around each joint. Select the different parts of the capsule manipulator to resize in different ways, using the Shift key to constrain in different ways.

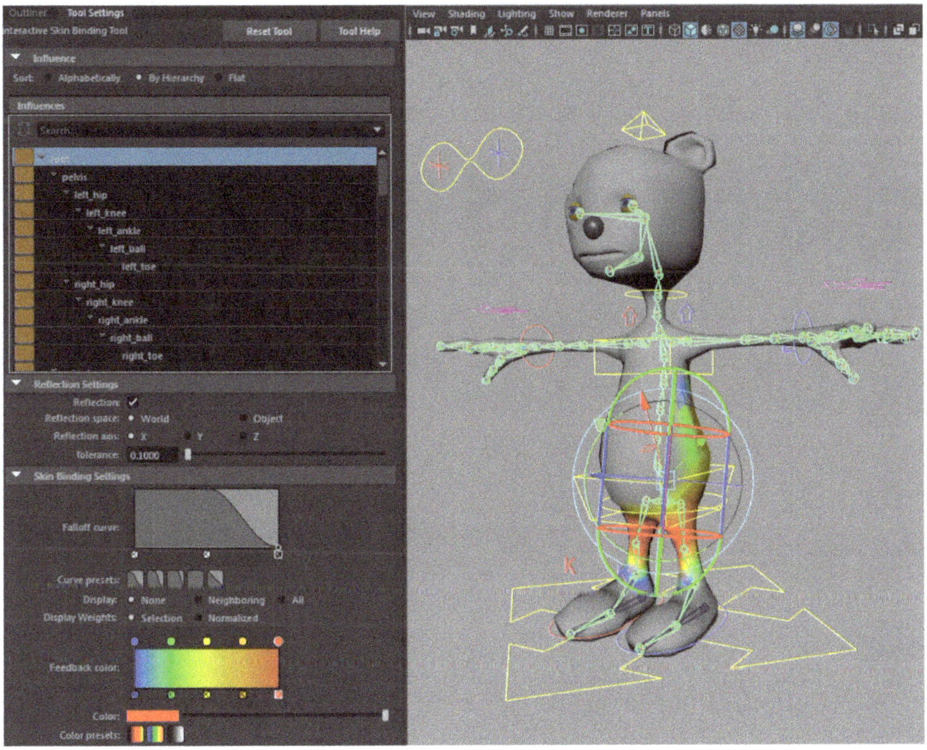

In general, capsules should be resized to overlap neighboring joints, but not by much! Also make sure that the capsules are wide enough to capture some of the vertices of the skin (especially in areas like the spine where the joints are far from the skin). At any point you can move the animation controls to see how things are deforming and return to Skin Binding with **Skin > Interactive Skin Bind Tool**.

TIP

It is helpful during skinning to first create a Technical Animation for your character, so that deformations can be viewed without leaving the Skin Binding and Weighting tools by moving to different animation poses on the timeline. To set this up, see *Part VI: Character Rig Testing.*

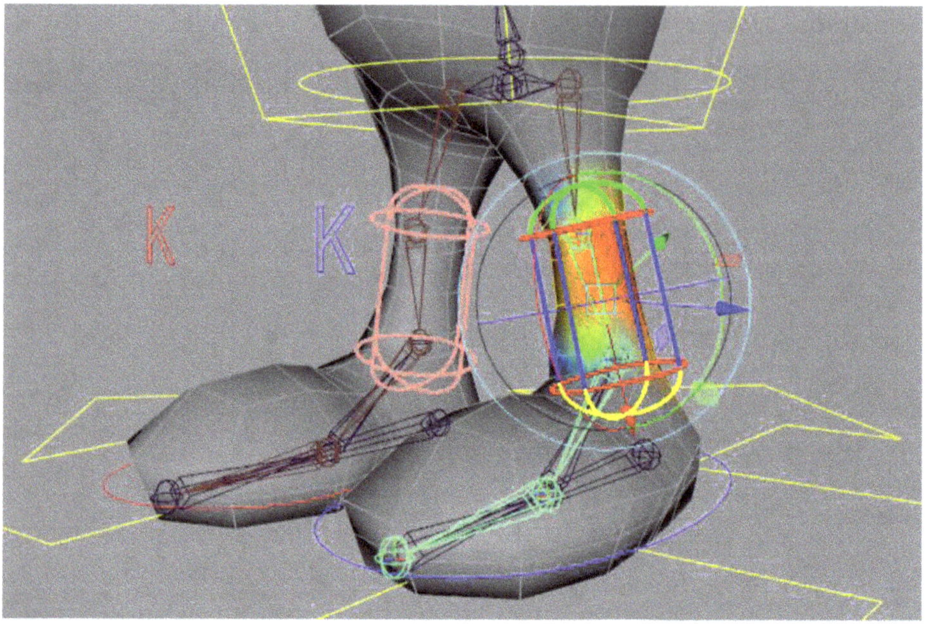

TIP

With **Reflection** turned on you only need to adjust capsules for one side of the rig.

Next, we'll bind the nose to a single joint. Select the nose geometry and the head joint and **Skin > Bind Skin (options)**: Change **Bind to** to **Selected joints** and **Bind Skin**.

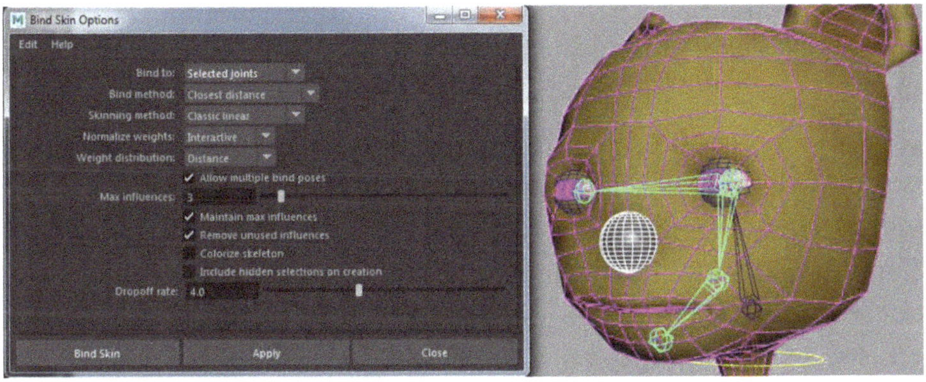

4.2 PAINTING SKIN WEIGHTS

To further refine the skin, select the geometry, make sure you are in **Shading > Smooth Shade All** mode and go to **Skin > Paint Skin Weights**. In the Tool Settings window, select joints on the **Influences** list and paint the geometry to adjust the area of influence. For quick results, paint areas of influence with **Paint Operation: Replace** and a value of **1** (a pressure-sensitive tablet will vary the value). Geometry will appear full white at a value of 1 and black at a value of 0. In the next pass, go back through with **Paint Operation: Smooth** and soften the transition areas. Concentrate on one half of the body, and then use **Skin > Mirror Skin Weights** to copy the edits to the other half.

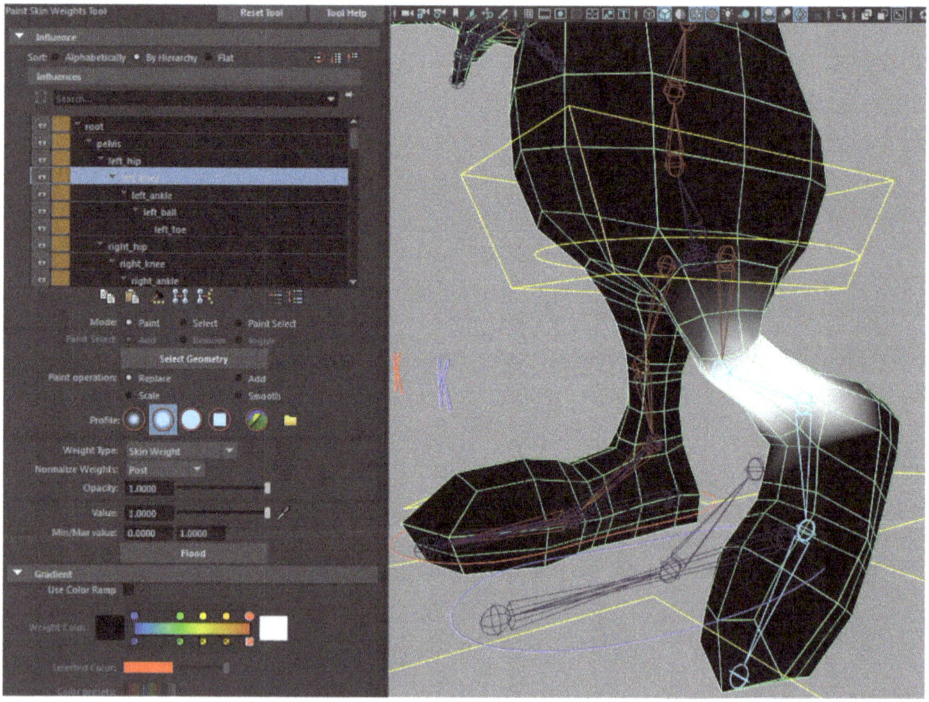

TIP

Click on a joint with MMB and use MMB to rotate the joints to test while using the Paint Skin Weights tool. Click with LMB to switch back to the brush. Alternatively, set keyframes for various calisthenics poses throughout your timeline and move to different frames to view the skin under different deformation conditions. There's also a handy RMB-menu to use over joints while using the Paint Skin Weights Tool. If you want to rotate joints while weighting, you can temporarily toggle off IK by using **Modify>Evaluate Nodes>IK Solvers**. As with all Maya paint tools, hold **B** and drag to adjust brush size.

The icons in the Paint Skin Weights tool settings window enable you to smooth selected vertex weights (Weight Hammer), copy and paste weights from vertex to vertex, and transfer vertices from influence to influence, all without leaving the tool! As a result of these updates to the tool, you will find less need to go into the Component Editor to adjust weights than in earlier versions of Maya.

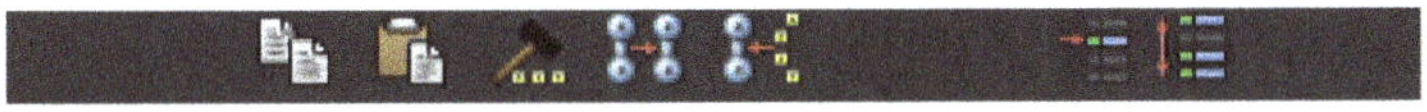

Below this row of tools is the **Mode** settings. For precision around joints, you can switch to **Select** mode, select edge loops on your geometry, select a joint from the influence list, and then manually type a value into the **Value** field for every vertex on that edge loop. (**Normalize Weights** should be set to **Interactive** during this workflow to keep the total weight value of all influences on a vertex normalized to a value of 1.) For example, you could select the edgeloop that cuts around the center of the knee, and type a value of 1 to give full weight to the knee joint. Then lock everything else on the influence list (the padlock icons next to the joint names) except for the two joints you want influencing the current selected vertices; in this case, *left_hip* and *left_knee*. Now if you select *left_knee* and update its value to **.5**, you are giving half its influence to the hip. Switch to *left_hip* and you'll see the values have auto-updated to **.5** as well (normalized to 1). For the rows of vertices above and below this row, use the same method to weight them **.7** toward the nearer joint (which will weight them **.3** toward the other joint). By working along edgeloops, you can achieve a nice ordered falloff of influence around joints.

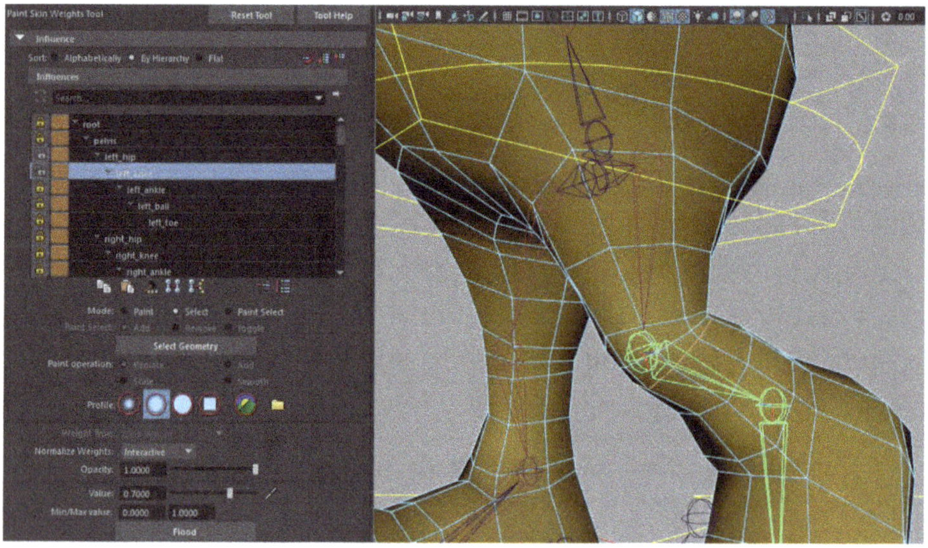

When you are all done working on joints, return their rotational values to 0 and toggle back on **Modify > Evaluate Nodes > IK Solvers.**

For the face, we're going to use blend shapes (morph targets) to drive animation, which we'll tackle in the next chapter. However, the best facial rigs use a layered approach utilizing both blend shapes and joint-driven animation. For this example character, we'll only be using a jaw joint, but it's the same basic process used in much more elaborate facial rigs, which frequently have joints along the lips and brows.

Select all vertices in the head starting above the neck and with the Paint Skin Weights tool, switch to **Paint** mode and select the *head* joint from the influences list. With **Paint operation: Replace** and **Value: 1**, hit the **Flood** button to weight all head vertices 100% to the *head* joint. (Note: you'll have to unlock all your influences for this to work.)

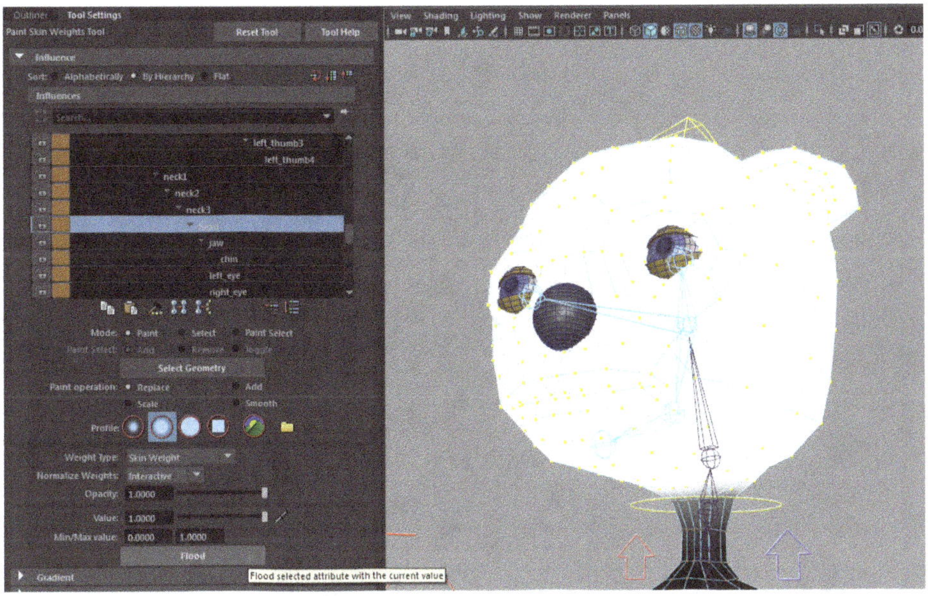

Then switch to the *jaw* joint and paint the area around the chin with a **Value** of **1**.

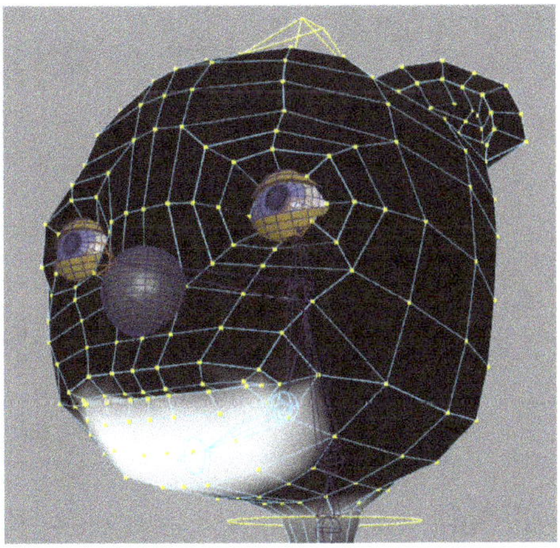

Rotate the *jaw* joint open to get to the vertices inside the lips and mouth sac. Switch back and forth between the *head* and *jaw* joints to fully separate the lower and upper jaw vertices. Lastly, switch to **Smooth** mode and with a very low **Opacity**, **Flood** a few times to smooth the transition area.

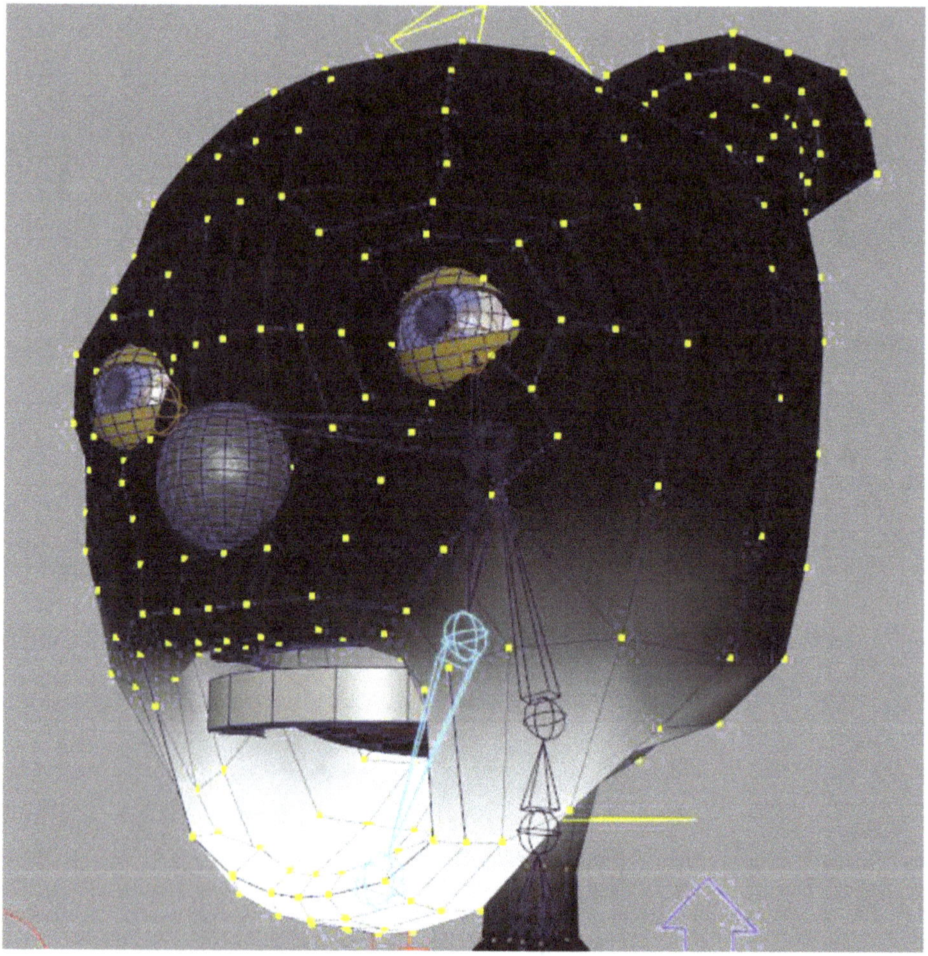

Keep testing every degree of rotation in the body and head and adjust skin weights. If a single vertex is out of line, select it and use the Weight Hammer tool (hammer icon below the Influences list).

When you are all done weighting the left side, don't forget to **Skin > Mirror Skin Weights**.

4.3 PAINTING DUAL QUATERNION BLEND WEIGHTS

Now go into the Attribute Editor for the *skinCluster* node on your main skin geometry and under **Smooth Skin Attributes** change **Skinning Method** to **Weight Blended.** This will allow you to use the Paint Skin Weights Tool to paint in areas of *dual quaternion* (DQ) skinning. Basically dual quaternion holds volume around joints better, so you will paint the shoulders, elbows and knees with dual quaternion. By leaving the other areas alone we default to the classic linear skinning method (which most skinning artists find has more predictable results in non-joint areas). Go back into the Paint Skin Weights tool and switch **Weight Type** to **DQ Blend Weight** and then paint the elbows, shoulders, and knees white (DQ) while leaving all else black (classic linear).

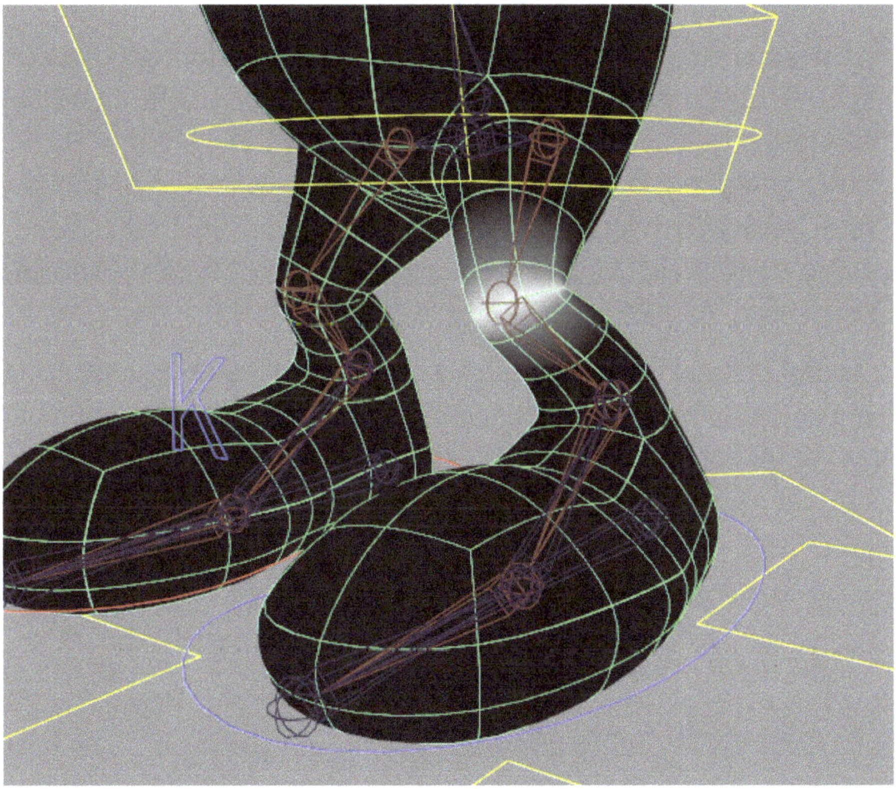

Continue refining the skin envelope, being careful to test every deformation as you go.

To clean up your skinning, finalize by going to **Skin > Prune Small Weights** and then **Skin > Edit Influences > Remove Unused Influences**.

We've just reviewed the standard skinning workflow for low-resolution characters. In a higher-resolution character, you might find that using the **Geodesic Voxel** bind method (rather than interactive binding) followed by a Delta Mush deformer (**Deform > Delta Mush**) can get you close to good skin deformation faster, but there's no substitute for weight painting for optimal results.

4.4 CORRECTIVE SHAPES WITH THE SHAPE EDITOR

The final step in skinning process is the creation of corrective shapes. These are blend shapes created to correct the inevitable deformation that happens in areas of extreme deformation, such as knees, shoulders, elbows, and hips. We'll look at two methods of creating corrective shapes; the first uses Blend Shapes driven by Set Driven Key (SDK) and the second uses Pose Space Deformations (PSD).

Blend Shape Deformers allow you to morph a base shape to the shape of one or more target shapes. This morphing can be automated by using the rotation of a joint to dial it in and out.

Take a look at the character's elbow. Chances are, the deformation of the skin is less than perfect when the elbow is bent at more than a 90-degree angle.

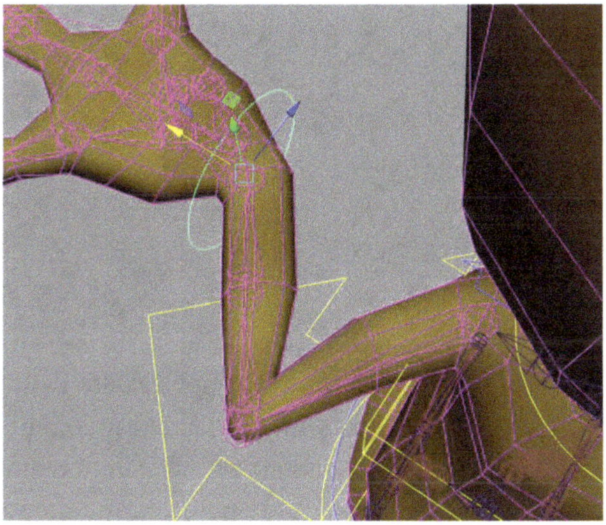

The first step in creating Blend Shapes is to create a Blend Shape Deformer, which can be thought of as an empty container to hold a collection of Blend Shape target shapes. With the skin geometry selected, go to **Windows > Animation Editors > Shape Editor** and hit the button to **Create Blend Shape**. Double-click the name of the deformer (***blendShape1***) to rename it ***correctiveShapes***.

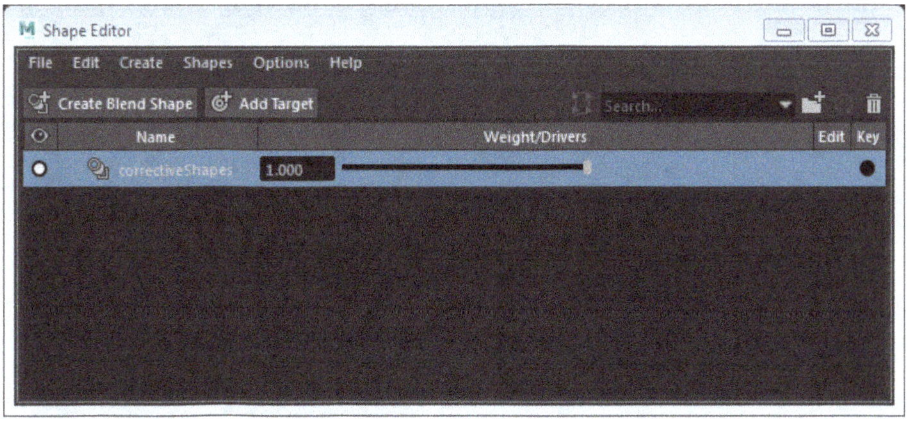

Next, hit the **Add Target** button and rename the target shape *l_elbow*. You'll notice the **Edit** button is highlighted red.

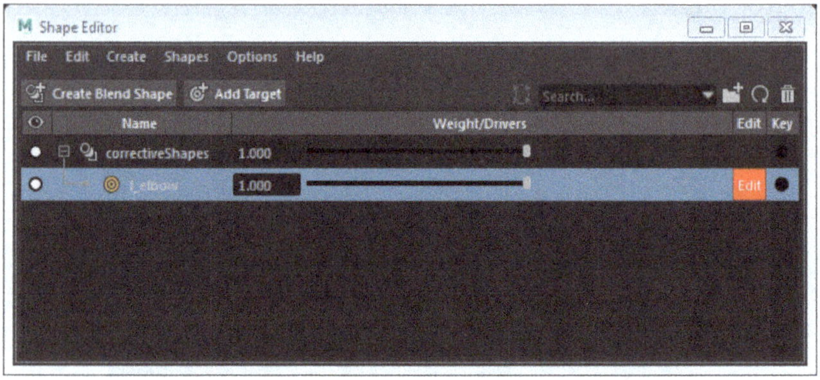

While in this Edit mode, you can re-sculpt the elbow to look better in this position. Any geometry edits you do in Edit mode will be applied to the target shape, not your original geometry. When done modeling, turn off the Edit mode button. Test your edits by dragging the weight slider.

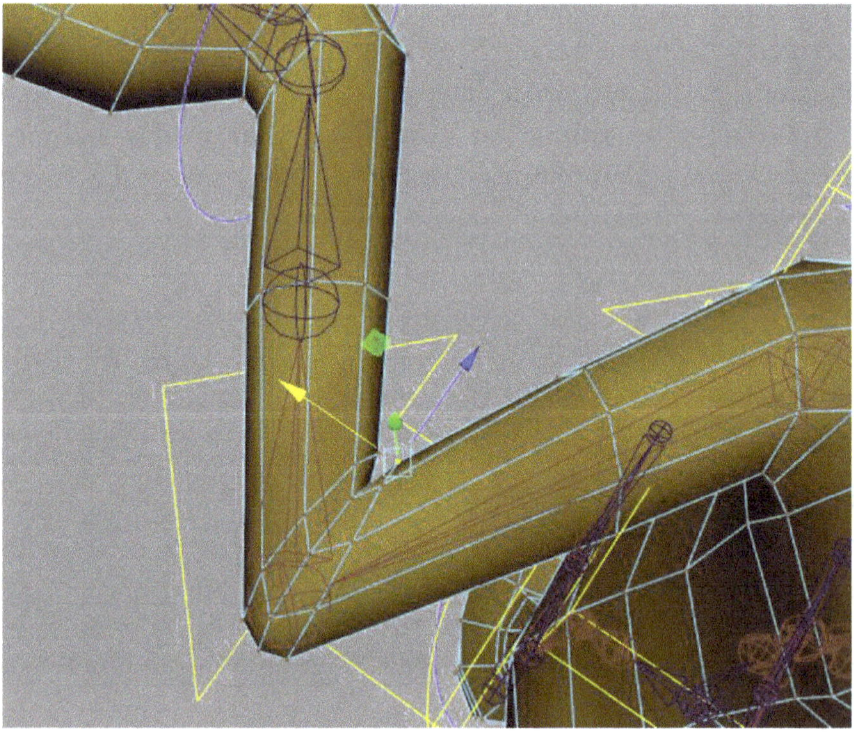

If you select the *left_elbow* joint, you'll notice that it has a Y rotational value of around -100 in this position. We'll use this to drive the weighting of the blend target. In the Shape Editor, rt-click over the value field of the *l_elbow* target shape and go to **Set Driven Key…**

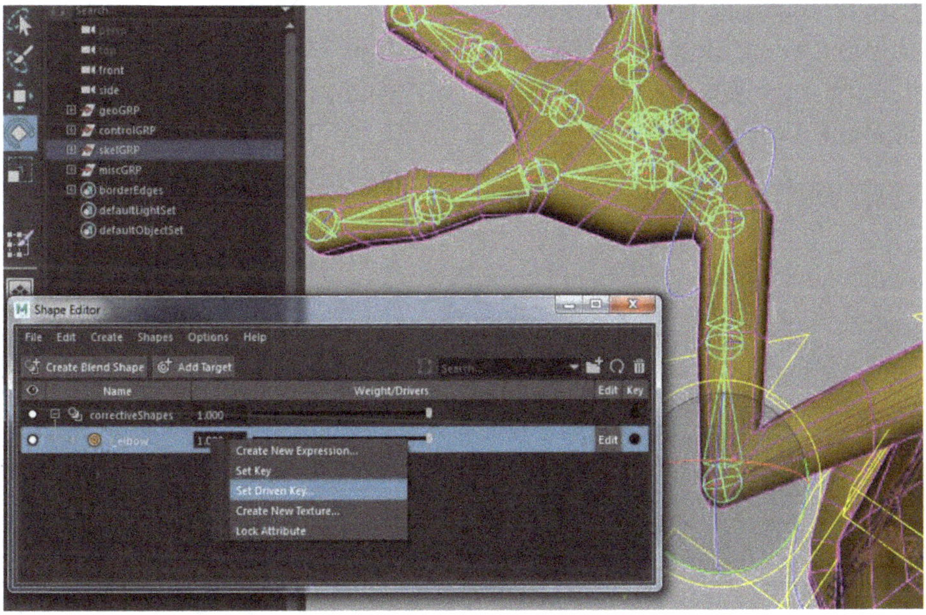

The SDK editor should come up with the *correctiveShapes l_elbow* already loaded as **Driven**. Select the *left_elbow* joint and click **Load Driver**. Highlight **Rotate Y** on its keyable attributes list.

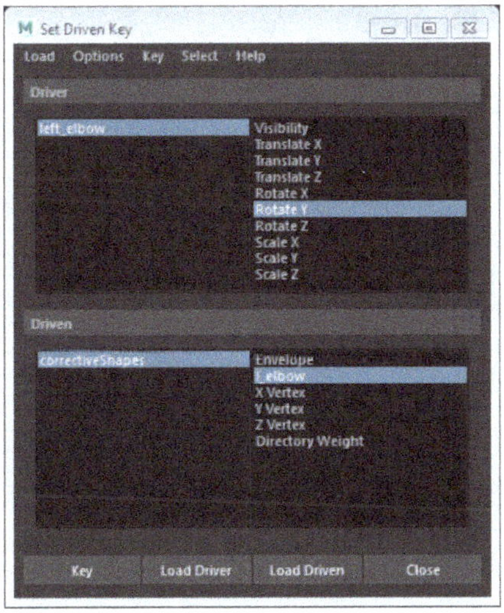

With the elbow still bent, hit **Key** in the SDK window. Then select the *left_armControl* and zero out its translate values (thus straightening the elbow). The arm looks ugly because the corrective shape still has full weighting. Drag the weight slider in the Shape Editor all the way to zero and **Key** the SDK again. You have now set up a relationship between the corrective shape and the angle of the elbow joint. Test by moving *left_armControl* and watch the weight slider in the Shape Editor automatically adjust for the angle of the elbow joint.

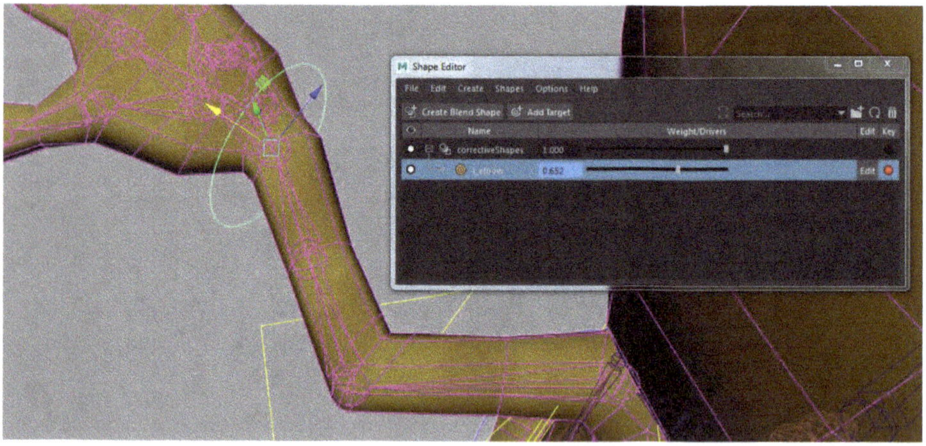

When you are happy with the deformation, you can mirror the corrective shape to the other elbow. In the Shape Editor, **Shapes > Duplicate Target** with the *l_elbow* target still highlighted. A copy of it will appear on the list. Rename it *r_elbow.*

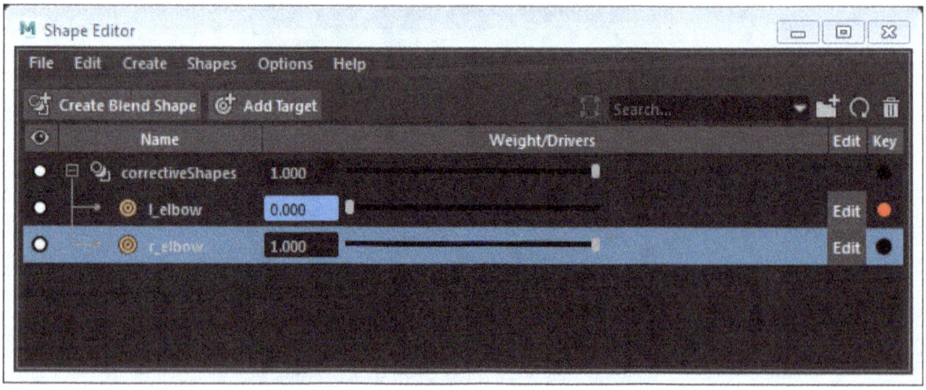

Next go to **Shapes > Flip Target,** which will flip the new corrective shape to the right side of the geometry. The SDK connections do not flip, so manually set SDK with this new corrective shape as you did with the left side.

4.5 CORRECTIVE SHAPES WITH POSE SPACE DEFORMATIONS

The shoulder poses a challenge because it is a 3-axis joint (as opposed to the elbow only rotating in the Y axis). So there is no single joint's rotational value that we can use to drive the SDK of the blend shape. Instead, we'll use a more advanced system of manipulating corrective shapes, Pose Space Deformations.

Windows > Animation Editors > Pose Editor. With the *left_shoulder* joint selected, click the **Create Pose Interpolator** button. The Pose Interpolator node calculates the joint's *pose* by evaluating the *twist* axis combined with the other two axes of rotation, called the *swing* axes. It stores a neutral pose for twist only, swing only, and twist and swing.

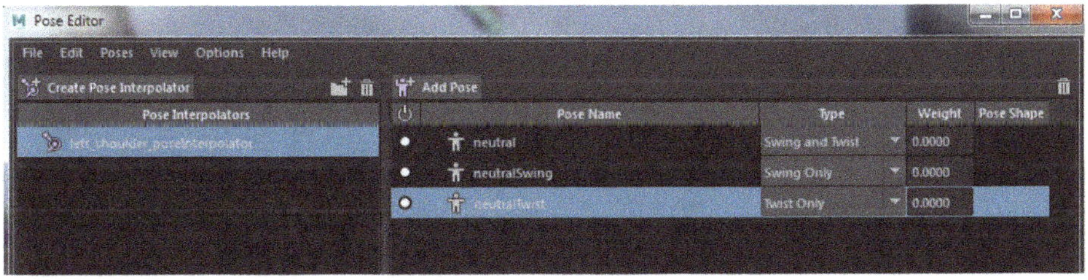

Rotate the *left_shoulder* joint in Z to put the character's arm down—easiest to do by temporarily turning off IK and rotating the joint directly: **Modify > Evaluate Nodes >** (uncheck) **IK Solvers**.

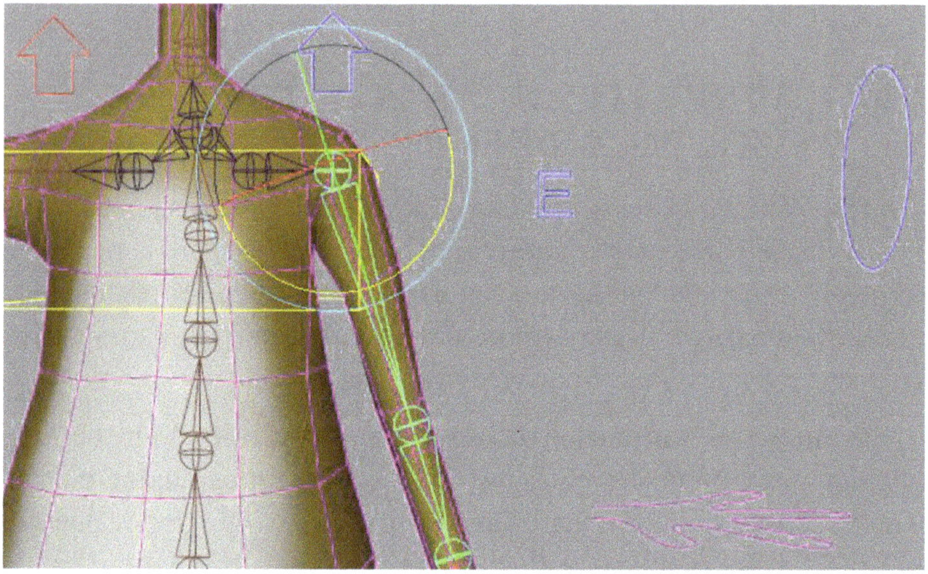

Now click the **Add Pose** button. In the dialog, keep **Create Pose Shape** checked and use the drop-down menu to add the new target shape this will create to the existing *correctiveShapes* Blend Shape Deformer.

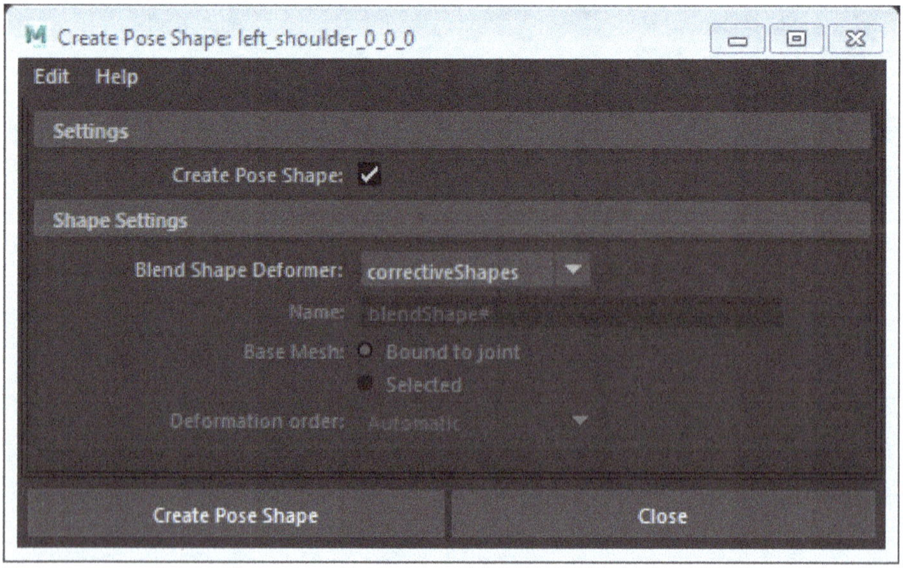

Now your Pose Editor contains a new pose with rotate values in its name—in my case, *left_shoulder_0_0_n72*, indicating this pose kicks in as the Z value nears negative 72.

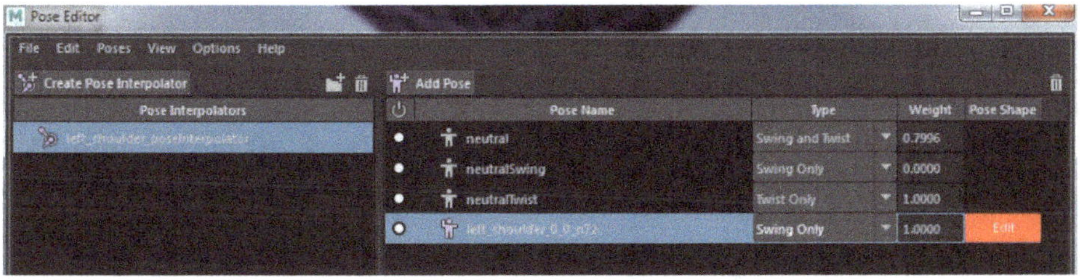

As we are already in **Edit** mode, you can use any modeling or sculpting tools to refine the shape of the shoulder at this pose. Don't forget to turn off **Edit** mode when done sculpting. Test by rotating the joint and observing the blend shape weight value changing.

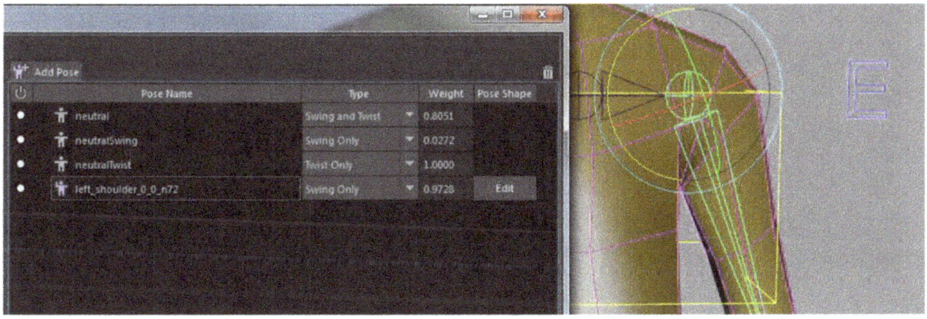

When done testing, you can rt-click over any of the neutral poses and choose **Go To Pose** to return to the neutral state.

Now test the other axes. The blend shape isn't triggered when rotating in X, as that is the twist axis. But the blend shape is triggered when rotating in Y, since it is a secondary swing axis. To turn off this potentially troublesome behavior, go to the Advanced Settings: **View > Advanced.** Here, observe that the twist axis is defined as the X Axis, so no trouble there, but both swing axes are triggering the blend shape. To fix this, select and **Go To** the corrective pose. Check **Independent** and under **Rotation Falloff**, enter the maximum rotation value for the pose (copy it from the name of the blend shape).

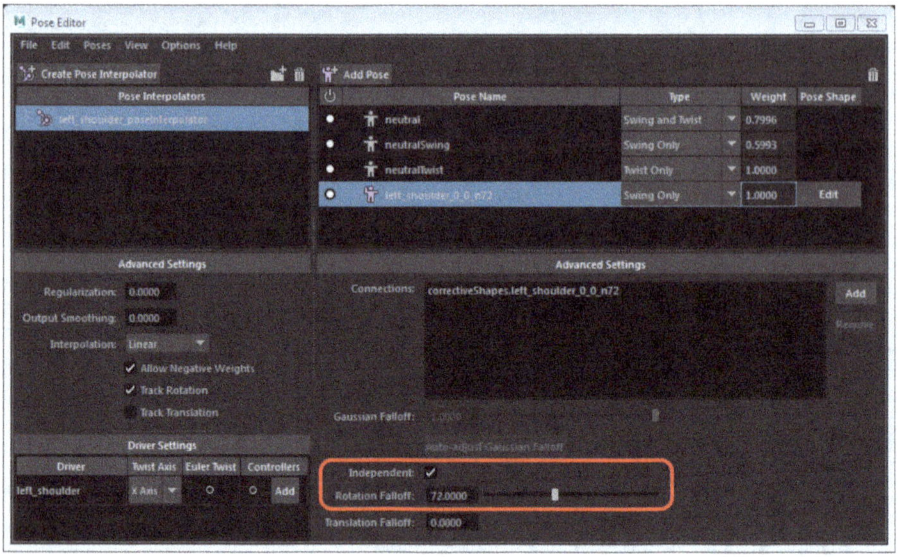

Now when you rotate in the Y axis, the weight of this blend shape doesn't update.

Now create another corrective shape for the Y axis: Swing the arm all the way forward in Y (about -90 degrees). **Add Pose** again, and this time we can go ahead and set it to **Independent** with the **Rotation Falloff** from the Y value in the new blend shape's name.

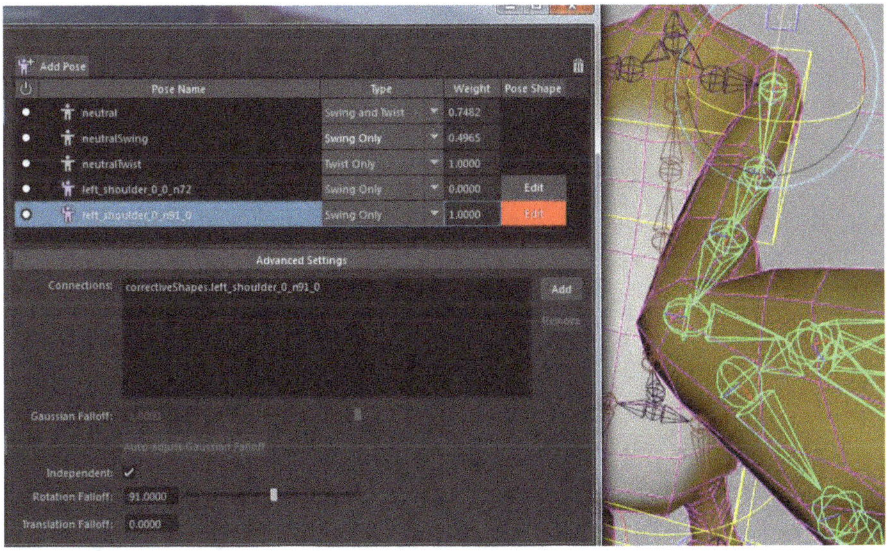

Proceed to edit the geometry around the shoulder to look good in this pose. Don't forget to turn off **Edit** mode when finished.

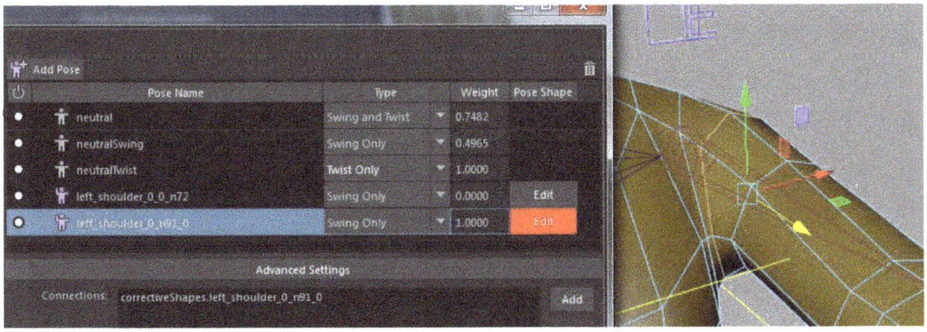

TIP

For editing shapes, we're just transforming vertices and edges, but for a more typical higher-resolution character the tools under **Mesh Tools > Sculpting Tools** are very useful, particularly those under **Shape Authoring**.

Test out your shoulder in a variety of poses to see how the corrective shapes work with each other.

When you are done with the left shoulder poses, select and rt-click over the name of the *left_shoulder_poseInterpolator* in the Pose Editor and choose **Mirror**. In the popup, set **Source** as ***left_*** and **Target** as ***right_***. This mirrors the interpolator and all associated poses to the right shoulder.

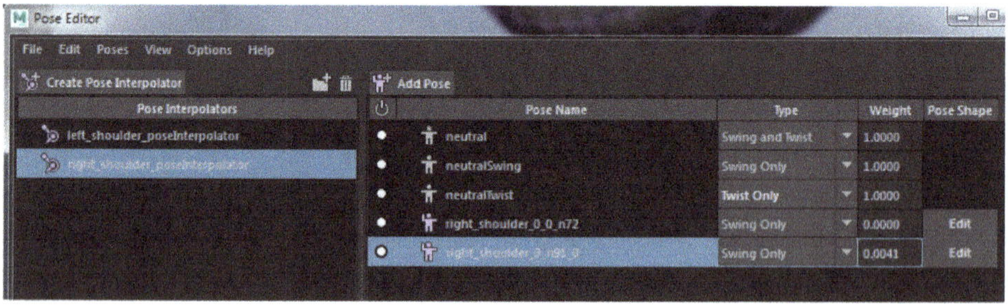

Enable IK and test both shoulders.

Try applying these same techniques to the knees and hips. When you are done, clean up the *poseInterpolator* nodes in the Outliner by moving them to the *miscGRP* which also includes IK handles and other miscellany.

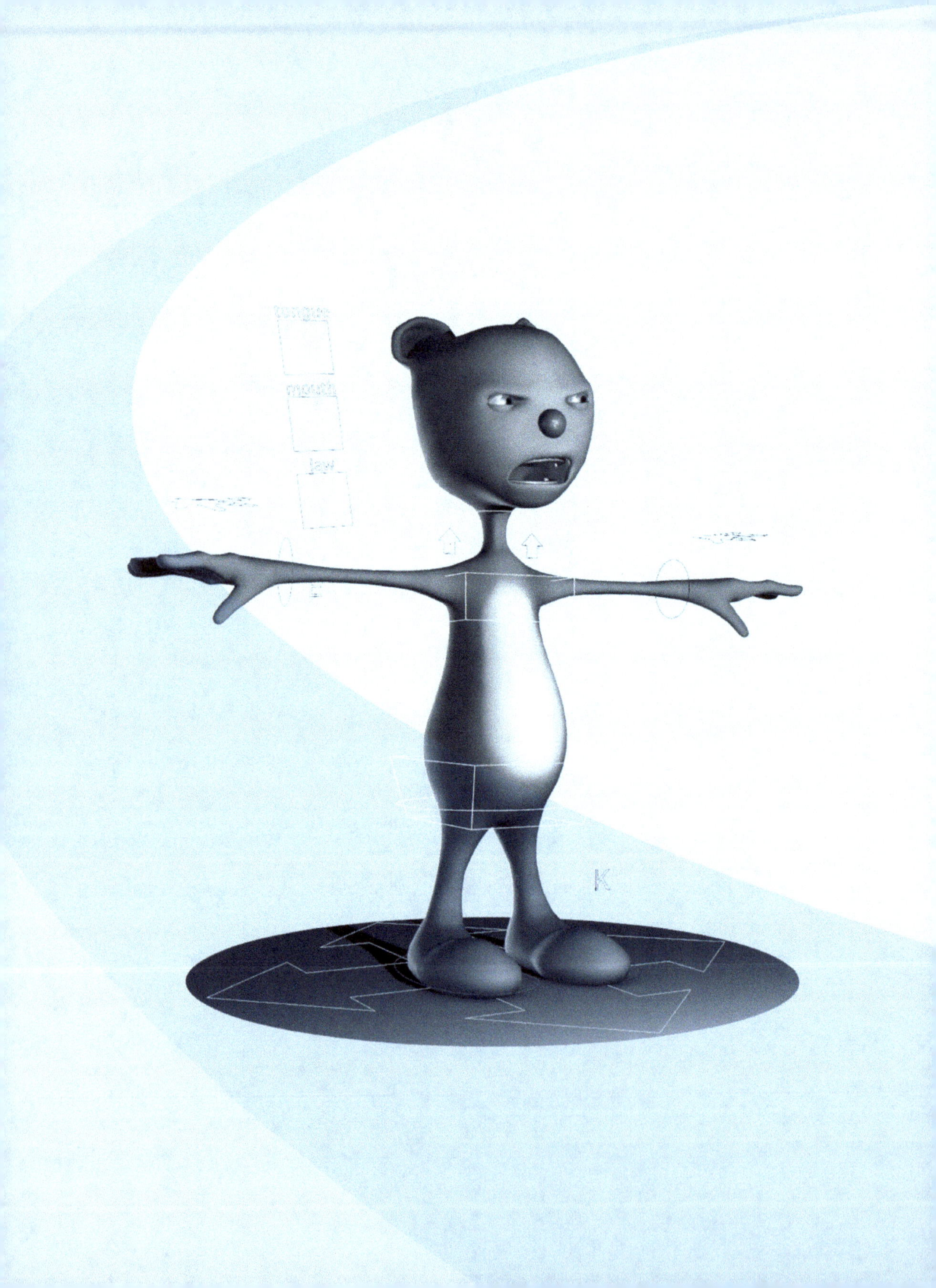
tongue
mouth
jaw

Part V
Character Facial Setup

5.1 SETTING UP FACE BLEND SHAPES

We've seen how blend shape deformers are used for corrective shapes in skinning. Another common use of blend shapes is as morph targets for facial expressions. Whereas the Shape Editor allows us to work on a single mesh in edit mode, facial expressions are typically done on separate duplicate meshes so that the inventory of face shapes can be viewed and evaluated in a side-by-side comparison. First we'll duplicate the entire character mesh for facial blend shapes.

TIP

Notes on head extraction. Because this is a seamless mesh, we'll copy the entire character. Only vertices that move will be evaluated in a blend shape, so duplicating the entire body mesh won't create too big a performance hit (although it will increase scene size). If you'd like to separate the head you can perform an extract operation, duplicate that for blend shapes and then combine the second mesh back to the body; however, this method will leave you with a mesh that you cannot delete history on, nor adjust skin weighting, so it is not recommended. The third option would be to download the popular *detatchSeparate* script from creative crash (*https://www.highend3d.com/maya/script/detachseparate-mel-for-maya*) which will allow you to extract the head with no history, leaving the body untouched. However, the head will have to be reskinned afterwards. You may wish to design characters with heads modeled as a separate mesh to begin with (easier with clothed characters), if you know you will be doing blend shapes for facial animation.

- If you haven't already, group all your meshes into a group called ***geoGRP***
- Duplicate the ***geoGRP*** group and move new one to the side (Ctrl+D) (By duplicating the group, we are copying the eyes and teeth as references so we can create brow and mouth blend shapes properly)
- Shift+D to duplicate with transform 5 more times
- Select those six, duplicate them, and move them up to be a second row (12 in all). The top six will be the upper face blends, the bottom six the lower face blends

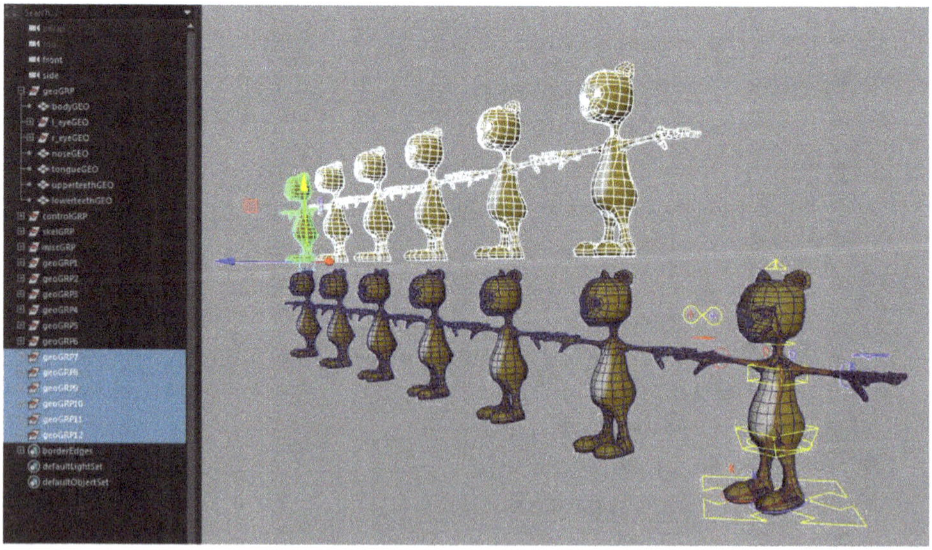

- Create the blend shapes (listed below) by using the move tool with soft select (**B**) or the sculpting tools. Concentrate on one side of the face because we'll duplicate and flip them to create the other side. Name each copied *bodyGEO* mesh the name you want the blend shape to be (*l_browUp*, *l_browDown*, etc). Be careful not to move any vertices but the ones in the local area of the particular blend shape. For example, brow shapes should only move vertices around the brow.

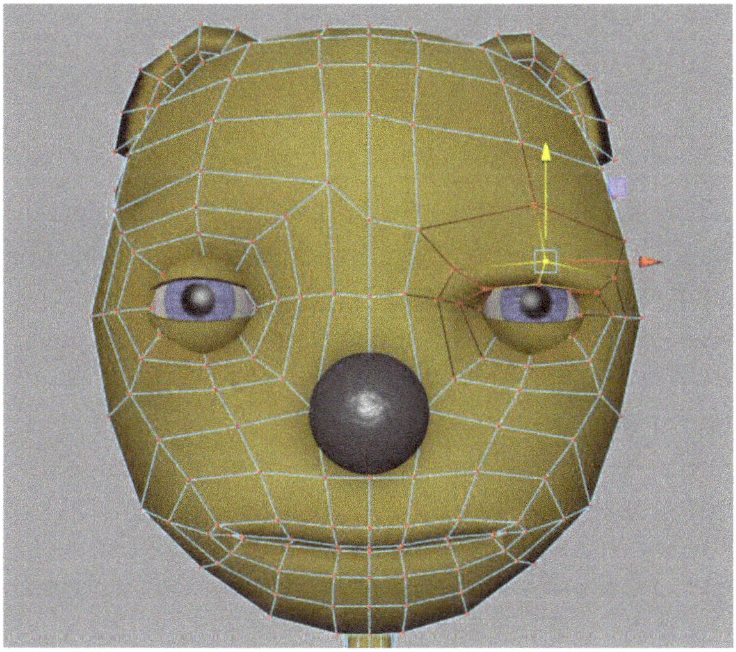

The Upper Face Shapes:

- Brow down—brow pushed down and slightly forward over eyes
- Brow up—"surprise" brow, pushed up and back along forehead
- Worried—inner brows raised, outer brows lowered
- Angry—outer brows raised, inner brows lowered

- Cheek squint—flesh mainly below eye scrunched up, concentrating (can be used in combination with Brow Down to make a squint)
- Cheek puff—cheek inflated as if blowing

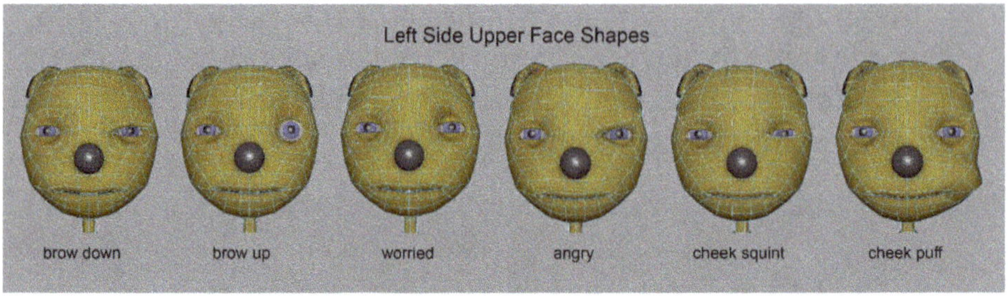

The Mouth Shapes:

- Mouth Narrow—lips pursed or puckered as if blowing or saying "oooo"
- Mouth Wide—lips long and thin, corners pulled back around the contours of the face
- FV—lower lip curled inward at the top, to fit under the upper teeth in a closed position (as in saying "f" or "v")
- Frown—ends of mouth pulled slightly downward
- Sneer—upper lip slightly curled outward and nasal folds lifted higher around nose
- Smile—ends of mouth pulled back and slightly upward

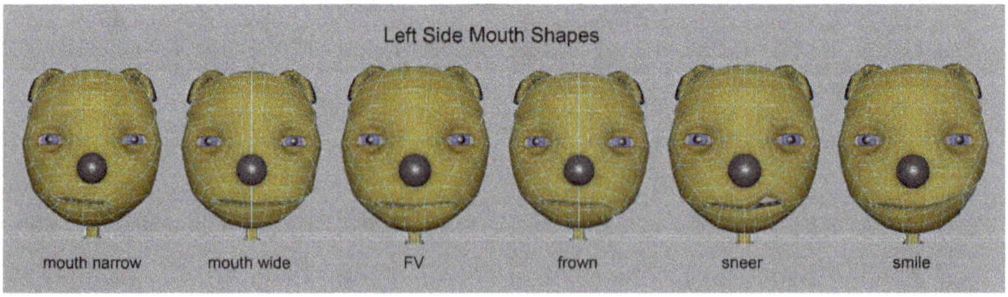

TIP

There are good online references for facial expressions. A good place to start is Pinterest, or look up Paul Ekman's Facial Action Coding System (FACS).

By making blend shapes very local they are as modular as possible; that is, meant to be used in combination to create the widest variety of possible shapes. Notice we didn't make mouth open/ mouth closed shapes; that's because we weighted the jaw area to be driven by a jaw bone, which we will animate for open/closed mouth action.

We will be mirroring shapes to the other side, so vertex movement along the centerline could double-transform when both sides are dialed in together. Thereby, minimize centerline movement, keeping vertices on the centerline at all costs, but where vertical movement is unavoidable (as in *Brows Down*) we can later partially mask the left and right blend shape along the centerline to fix the problem.

- Shift-select each of the mesh copies, Shift-select the original (base) model last, and perform **Deform > Blend Shape (options)**

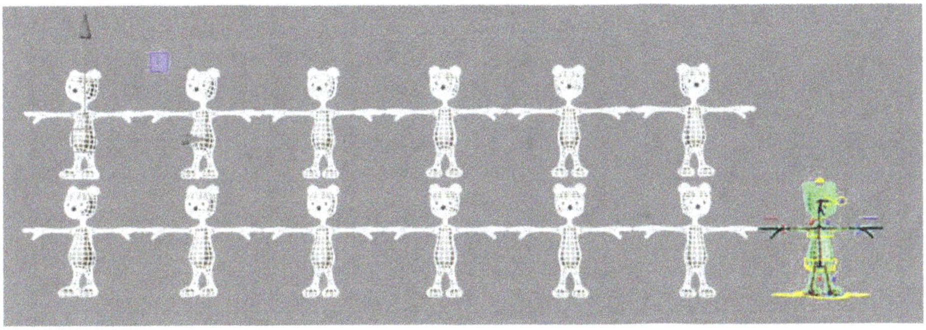

- On the basic tab, give it a name under **BlendShape node:** ***face-Blends*** On the **Advanced** tab, change **Deformation Order** to **Pre-deformation** (the blend shape deformer will evaluate before the skincluster). Hit **Create**.
- Open **Window > Animation Editors > Shape Editor** to access sliders for these new blends, or use the channel box (with base shape selected).
- You can now delete or hide all the target blend heads (you can change the targets to update the base head, so you may wish to

keep them around on a hidden layer, or you can always rebuild them from **Shapes > Rebuild Target**).

- Now we'll group and mirror. Select all the face blends in the Shape Editor and group them (Ctrl+G). Name this group *l_targets.* With this group selected, **Shapes>Duplicate Target** (Ctrl+D). Name this new group *r_targets.* With this new group selected, **Shapes > Flip Target**. Now rename all new targets from *l_* to *r_*. Try them out!

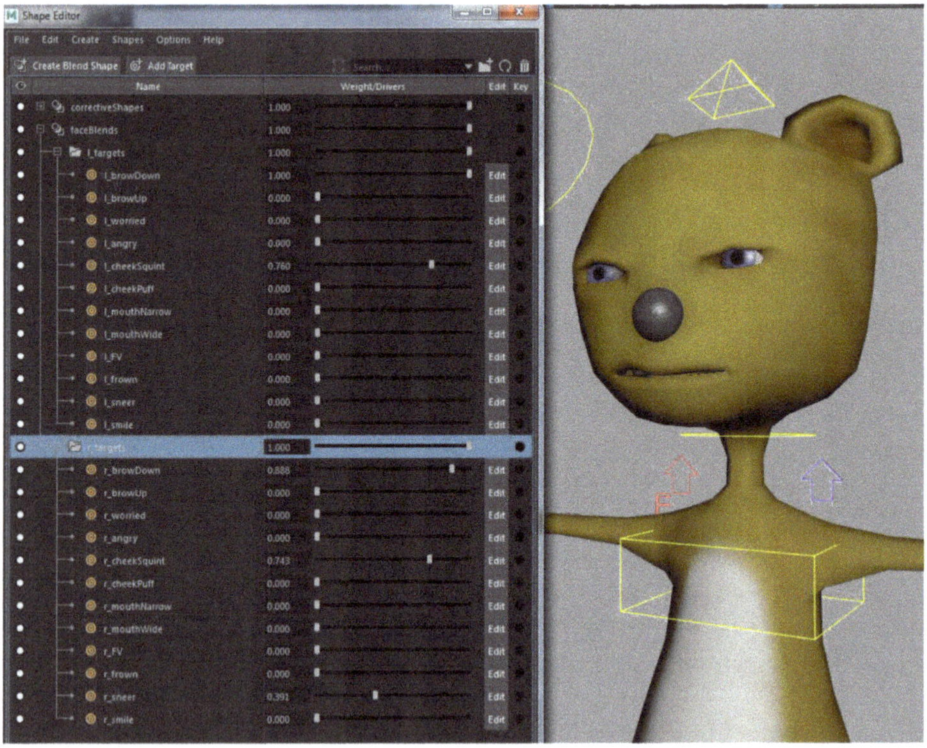

TIP

You can arrange these blend shapes into groups as you see fit; for example, putting the mirrored pairs together into groups. These groups have weight sliders that can be used to adjust them together. You can also select a left shape and use **Shapes > Mirror Target** to copy its edits to the other side for a unified symmetrical blend shape, but I find these less useful, as asymmetry is usually desired in animation. If flipped or mirrored pairs cause double transformation of vertices along the center line, you can partially mask the influence of the target shapes along the centerline using **Mesh Tools > Sculpting Tools > Mask Target Tool** and, with low opacity, paint along the center vertices.

5.2 SETTING UP EYELID CONTROLS

So now we have the body rig and facial blend shapes set up, but we haven't attached the eyelids or the teeth and tongue yet. Let's do that now.

First, the eyelids. Select the top and bottom left eyelids and Ctrl+G to group them. Name this group *l_eyelidGRP* and **Modify > Center Pivot**. Select the *head* joint and then the *l_eyelidGRP*. **Constrain > Parent** (maintain offset). Do this for the right eyelids as well. Test your rig by rotating the neck.

If you find rotating the neck in Z creates a weird googly eye effect (since the *eyesControl* isn't rotating with the head), select the head control, then the *eyesControl*, **Constrain > Orient (options)** and choose only the Z constraint axis. Then you can lock and hide Rotate Z for the *eyesControl*.

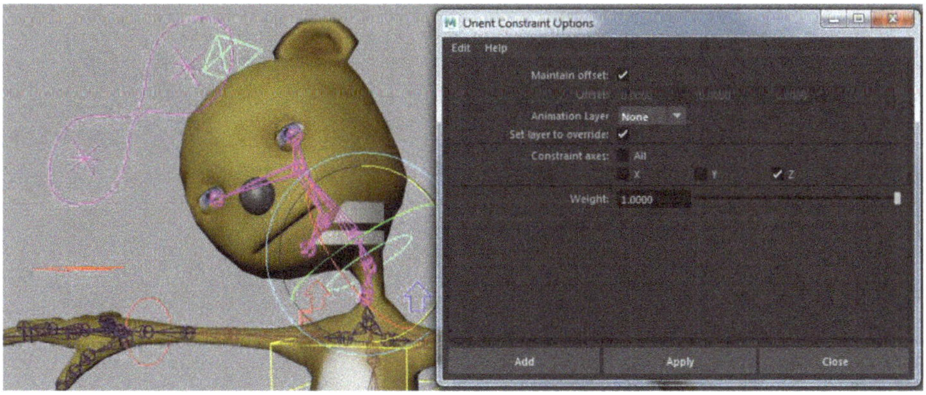

Now we'll use SDK to create blink sliders for the eyes. We'll add this attribute to the *eyesControl* so that all eye animation including eyelid rotation can be done with this one control.

Select the *eyesControl* and **Modify > Add Attribute…** Name it *leftUpperLid*, with **Numeric Attribute Properties** set to Min 0, Max 10. Add and then do the same to add *rightUpperLid*, *leftLowerLid*, and *rightLowerLid*.

(Animation Menu Set) Key > Set Driven Key > Set… Select *eyesControl* and **Load Driver**. Highlight **Left Upper Lid** in the attribute list and make sure it's set to 0 (channel box). Select *l_topLidGEO* (the upper lid) and **Load Driven**. Highlight **Rotate X** in the attribute list and set it to 0. **Key**.

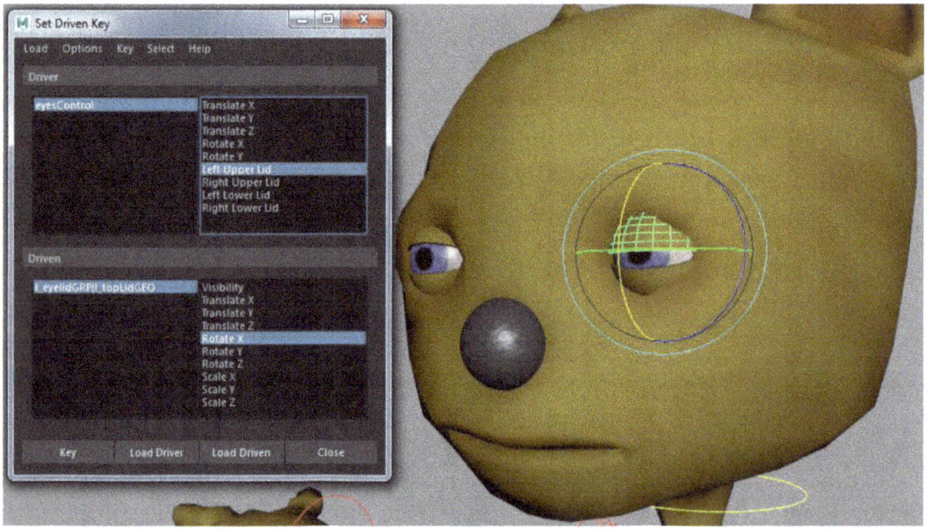

Now change *eyesControl.leftUpperLid* to 10 and set the *topLidGeo.RotateX* wide open until the eyelid just disappears into the face geometry—for me it's -45. **Key** and then test your **Left Upper Lid** attribute. When you are done testing, leave it at a comfortable value, like 5 or 6.

Do this for all four eyelids. As an option, you can also set up a *Blink* attribute on the *eyesControl* which drives all four lid attributes for ease of use when animating.

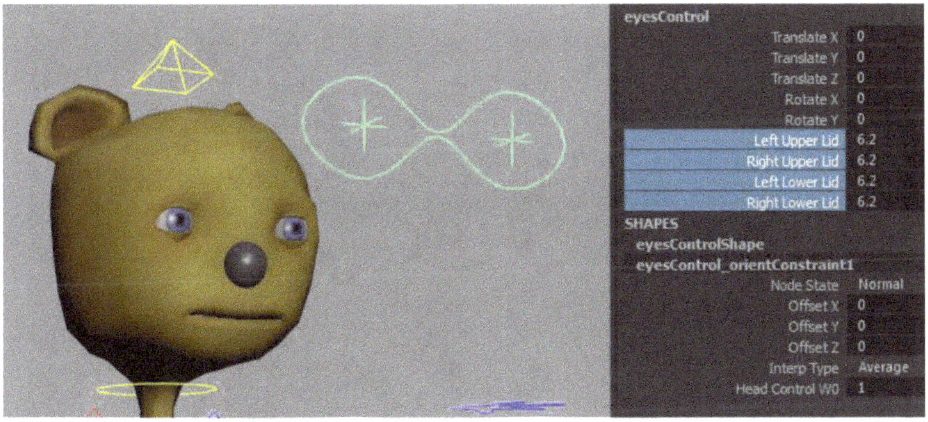

5.3 SETTING UP TEETH AND TONGUE

We've almost got this little guy animatable, but if you move him you'll notice his teeth and tongue still stay behind. We'll attach them to the rig next.

Parent constrain the upper teeth to the *head* joint. Parent constrain the lower teeth to the *jaw* joint. Test by rotating the jaw joint. You may have to remove the constraint and reposition the teeth if they are disappearing in the mouth sack or interpenetrating the jaw or neck.

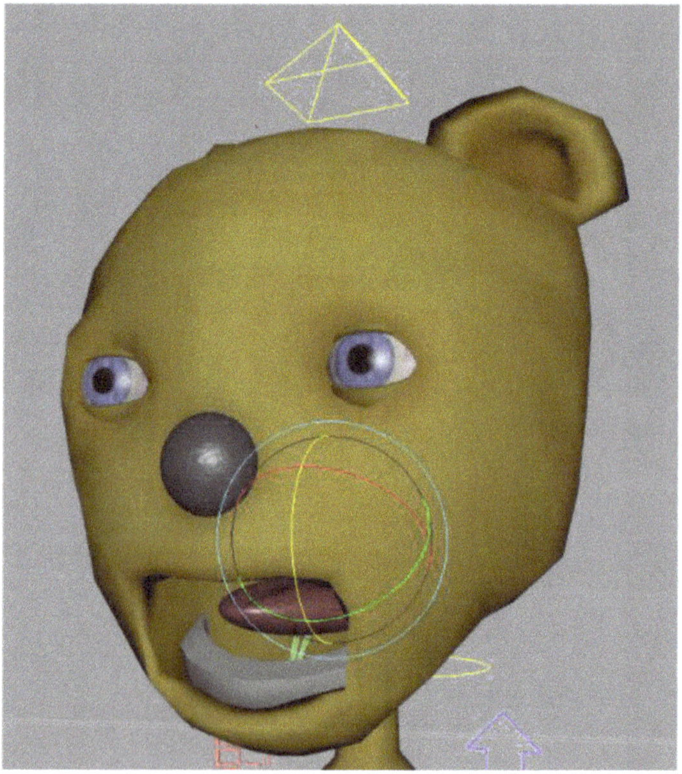

For the tongue, switch to the side view and **Skeleton > Create Joints** to draw a chain of joints from the base to the tip of the tongue geometry. Enter to complete the tool and then right-click over any joint in the chain to **Select Hierarchy**. Switch to the rotate tool and test rotating in Z to move the tongue up and down. They should all rotate together.

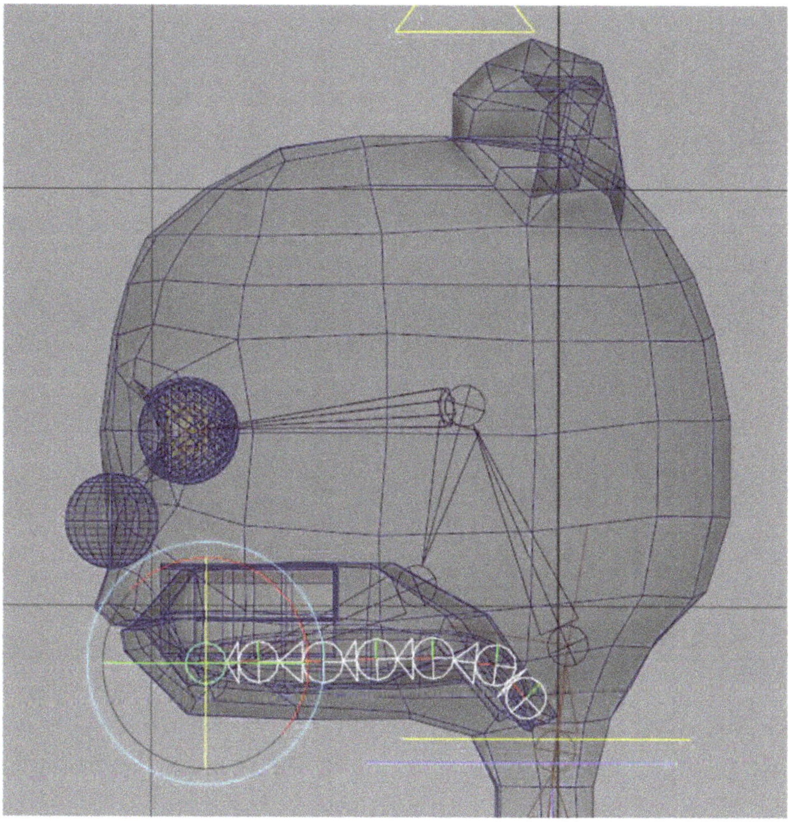

With the tongue joints still selected, go to **Modify > Prefix Hierarchy Names…** and add "tongue" as a prefix to the joint names. Then Shift-click the *tongueGEO* (may be easier to Ctrl-click it in the Outliner) and **Skin > Bind Skin (options)**. In the options, change **Bind Method** to **Geodesic Voxel** and then **Bind Skin**.

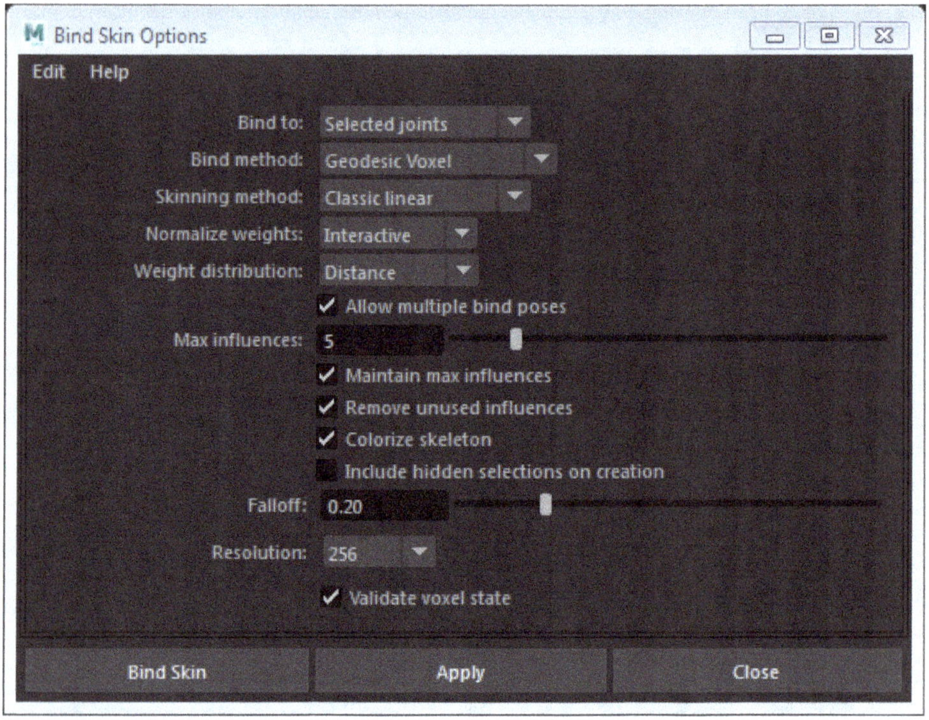

Test the tongue again. There should be no skinning issues. Select the base tongue joint (*tonguejoint1*) and Shift-select the *jaw* joint and hit **P** to parent. Now the tongue will follow jaw rotation. We'll set up SDK to create simple tongue lift and tongue roll actions, but for a more advanced character you'd want to add an IK Spline Handle with IK controls, similar to how we set up the back.

5.4 SETTING UP A GUI FOR FACIAL ANIMATION

Next we'll create control objects that animators can use to manipulate the jaw joint and the face blends. We'll do a couple of control objects together so you'll be able to see how this process works and then you can use the same techniques to build however elaborate a facial control GUI you want.

Using the **EP Curve Tool** set to **1 Linear**, create a square by grid-snapping (**X**) in the front camera view. **Modify > Center Pivot**. Move and scale this into place next to your character's head and freeze transformations. Name it ***mouthControlBox***.

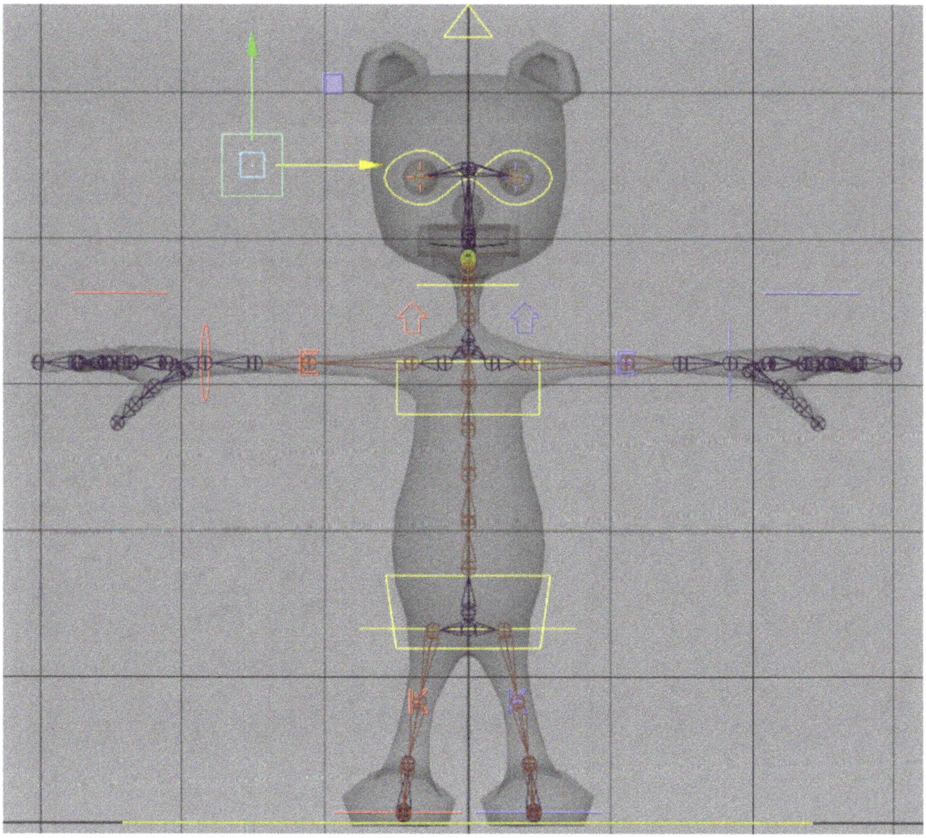

Create > Locator and point constrain it (**maintain offset** turned **off**) to the NURBS square, then delete the constraint. This is to place the locator in the dead center of the square. Freeze transformations on the locator. Name it ***mouthControl***. Parent it under the ***mouthControlBox***. In the attribute editor for *mouthControlBoxShape* node, under **Object Display**, turn on **Template** so the box itself can't be selected. You can turn this back off if you ever need to move the whole control. Also, in the locator's *mouthControlShape* node attribute editor, **Object Display** section, adjust the **Local Scale** in x, y, z so the locator itself is about 1/4th the size of the *mouthControlBox*. In my case, I set the values to **0.1, 0.1, 0.1**.

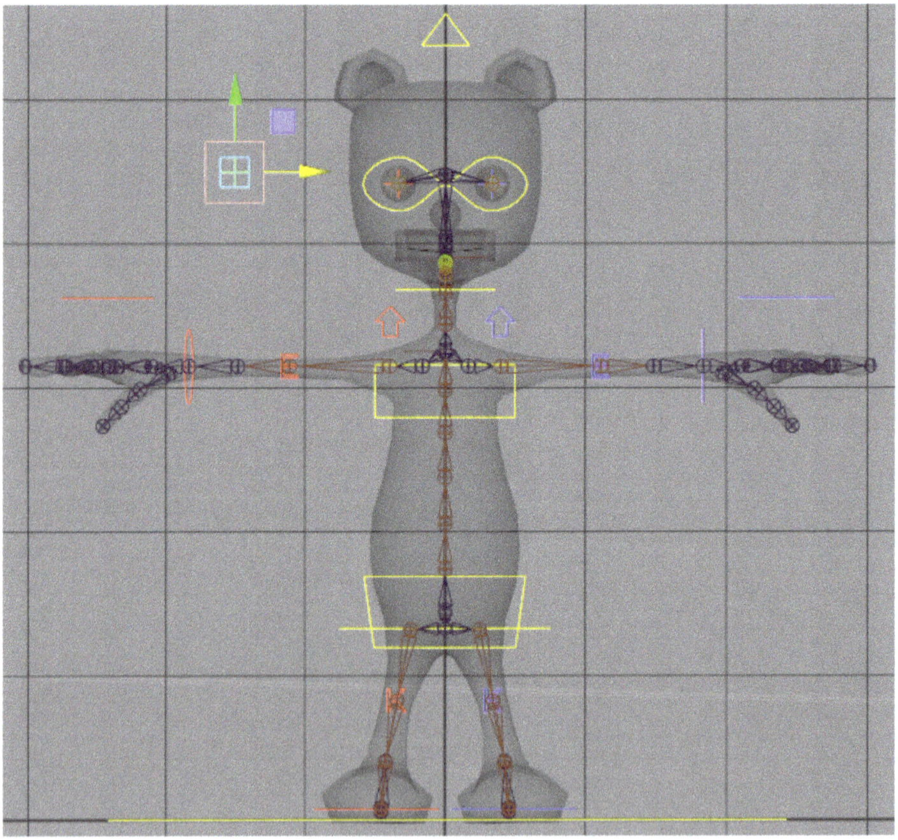

Now select the locator and move it in x to the edge of the square, taking note of what value you reach in the channel box. In my case, I created a 0.4 x 0.4 unit square, so the locator travels **0.2** units to reach the sides. This is where we will set the limits on the locator.

Go to the Attribute Editor for the locator and under **Limit Information** turn on limits for **Translate > Trans Limit** X and Y, setting **Min** and **Max** to **-0.2** and **0.2** (or your numbers noted in the previous step). Lock and hide *Translate Z*, *Scale*, *Rotate* and *Visibility* attributes. You now have a locator constrained within the limits of the NURBS square. Try it out.

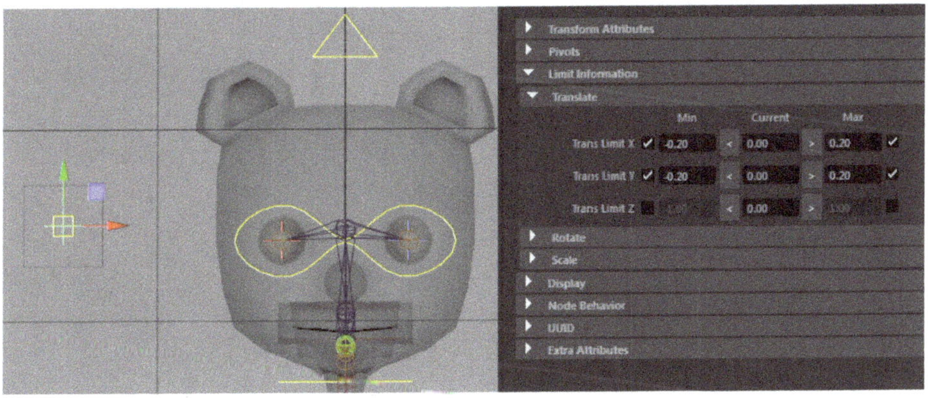

Now we'll use Set Driven Key to drive four blend shapes from one control object, simplifying the control system for the animator by consolidating several channels into one keyframe-able object.

Key > Set Driven Key > Set...

With the ***mouthControl*** locator still selected, hit **Load Driver**. Select the mesh and in the channel box select the ***faceBlends*** blend node. Hit **Load Driven**.

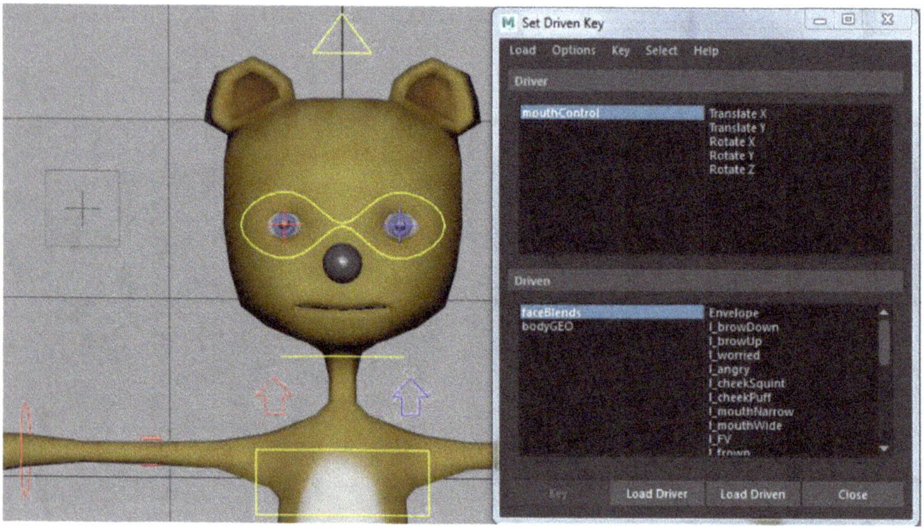

The *translateX* of the *mouthControl* will be used to drive the narrow and wide blend shapes. Select **TranslateX** in the upper field and **l_mouthNarrow**, **r_mouthNarrow**, **l_mouthWide**, and **r_mouthWide** in the lower field to set a default key at a value of **0**. Hit the **Key** button.

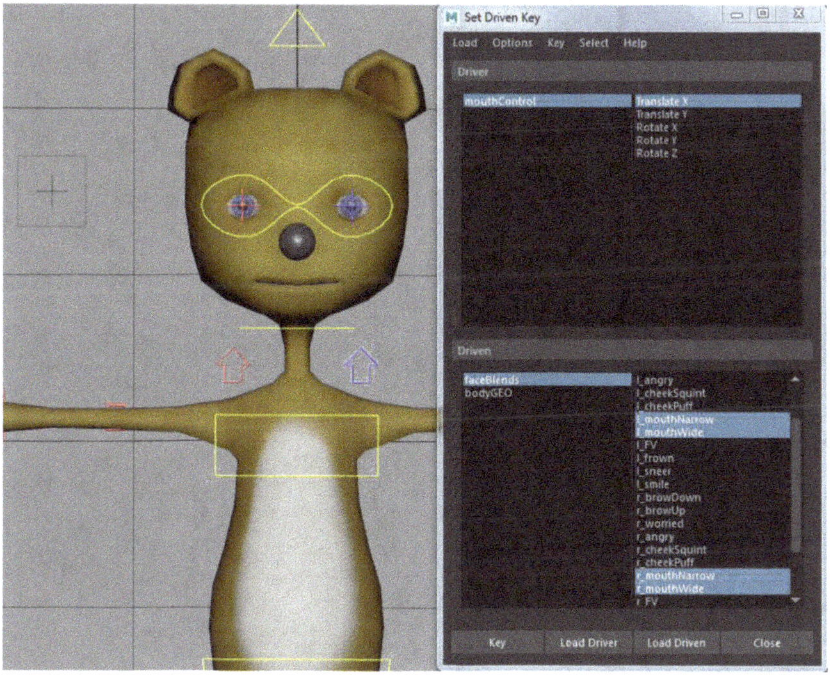

Now select ***mouthControl*** in the SDK window to bring its attributes into the channel box. Set *translateX* to the maximum: **0.2**. Select ***faceBlends*** in the SDK window to bring its attributes into the channel box and set the value for both ***Wides*** to their maximum: **1**. Make sure both are highlighted and hit **Key**.

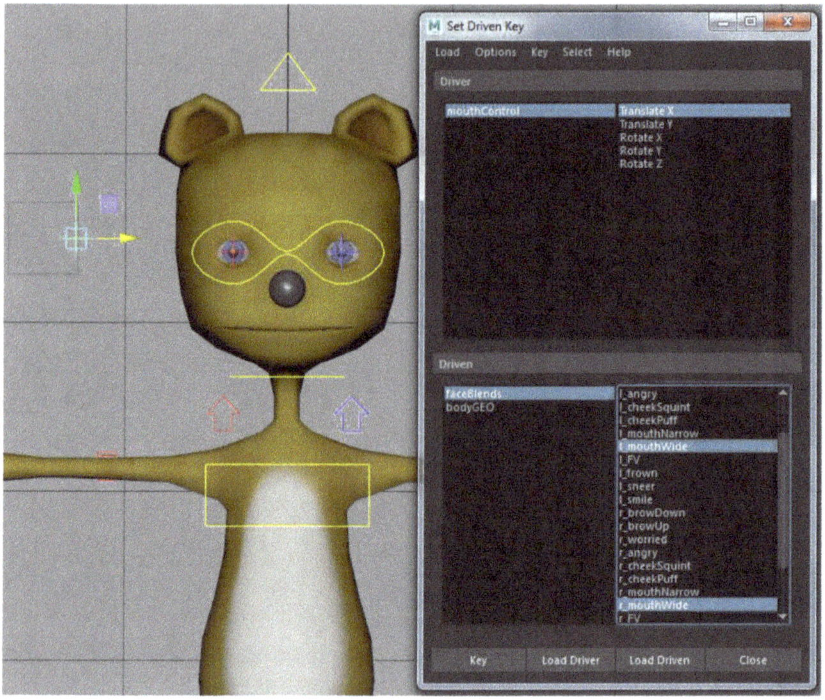

Now change the ***mouthControl.translateX*** value to its minimum: **-0.2** and the ***faceBlend narrow*** shapes to their maximum: **1** and hit **Key**. Test your slider by moving the *mouthControl's* translateX between its minimum and maximum to see it slide between the ***Narrow*** and ***Wide*** shapes.

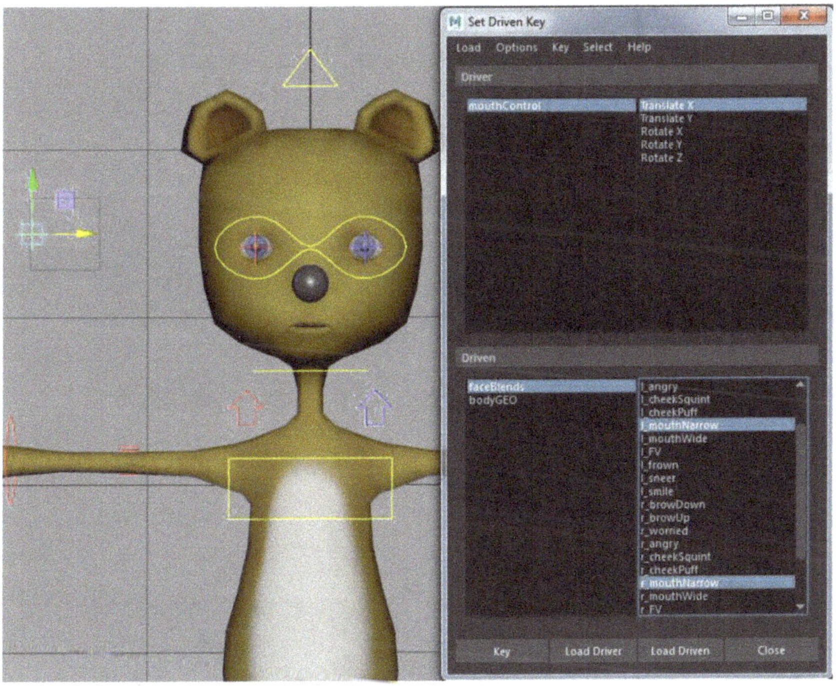

Now set a default 0 key on *mouthControl*'s **translateY** and *faceBlends*'s ***Frown*** and ***Smile*** shapes.

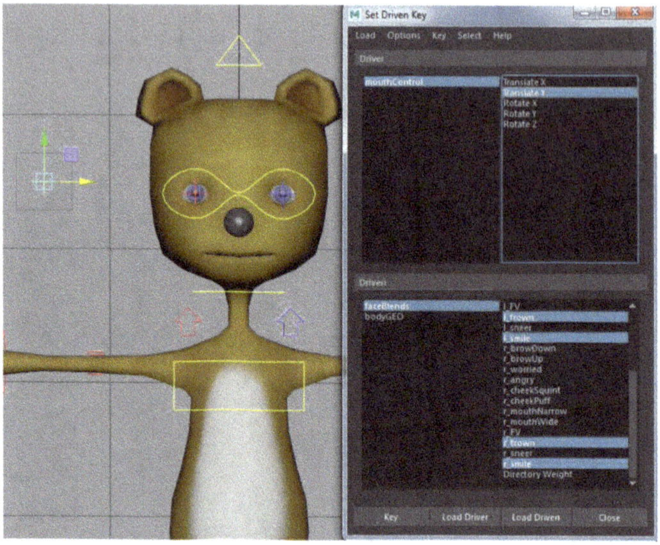

Set a Key with *mouthControl.translateY* set to its maximum: **0.2** and both *faceBlend.smile* attributes set to their maximum: **1**.

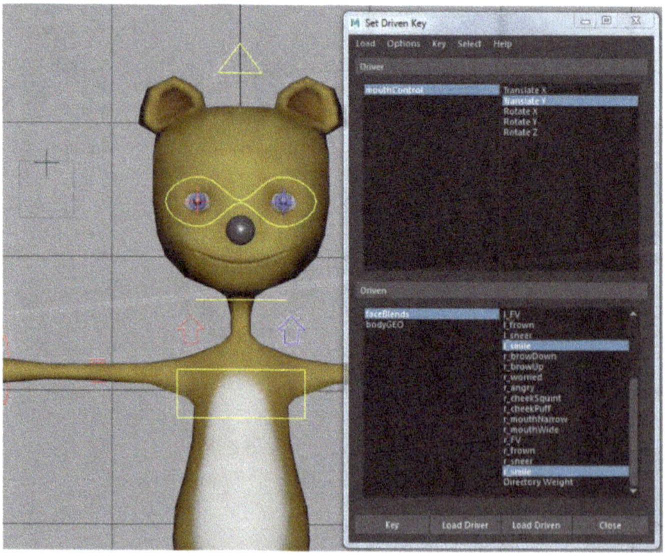

Now set a key with *mouthControl.translateY* at its minimum: **-0.2** and both *faceBlends.frown* attributes at their maximum: **1**.

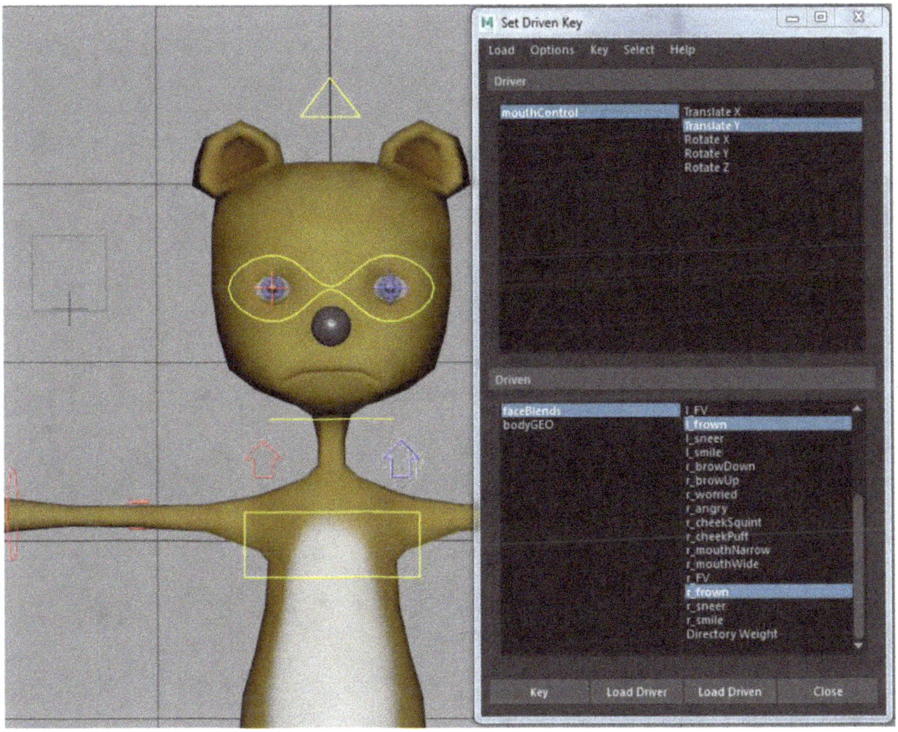

You can use **Create > Type** to create a label for the control, positioning it over the control box and templating it. (After you **Create Curves from Type** delete the *type1* node and size and position the curves.)

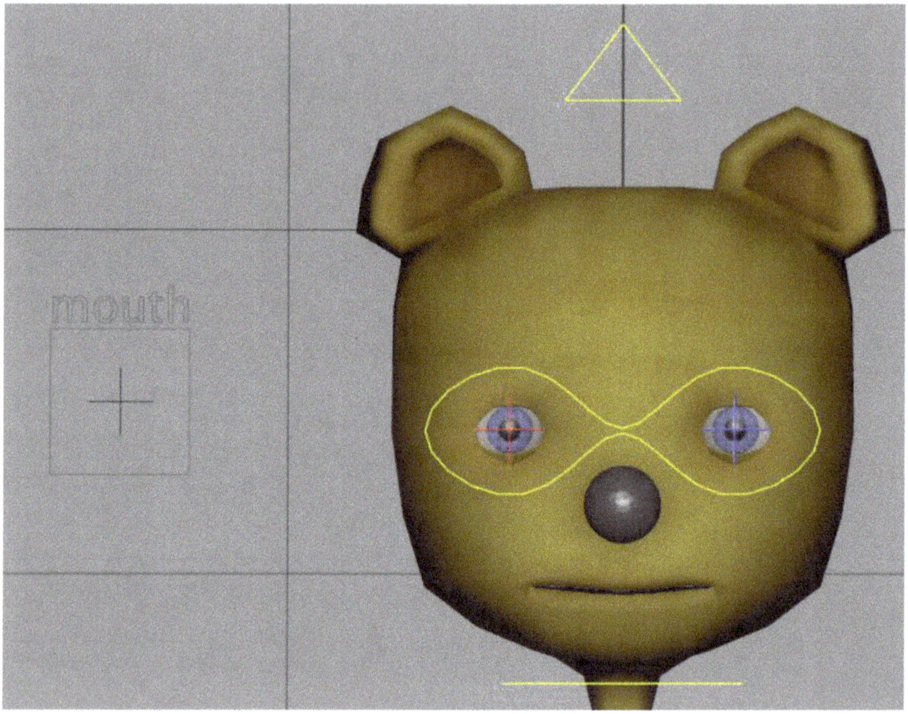

Now select the locator in your window and test your fully functional control slider. Pulling it to the top of the box invokes a smile, toward the bottom for a frown, to the right for wide shapes and to the left for narrow shapes. The corners blend between harmonious shapes (narrow smile, wide frown, etc.) but the overall setup prohibits the simultaneous invoking of shapes that are in opposition to each other. For example, simultaneously invoking both narrow and wide shapes would produce strange results, since the vertices are moving in opposite directions, so you would never use these incompatible shapes simultaneously. This setup prohibits such awkward distortions and makes the blend shape system easy for the animator to control and key.

Parent the type curves under the *mouthControlBox*, then group the *mouthControlBox* and name the group ***faceControls***. Parent this group under *headControl*.

Now we'll use a similar procedure to create a jaw joint controller. Select *mouthControlBox* in the Outliner and Ctrl+D to duplicate and move it to the side. Rename this *jawControlBox* and the locator underneath it *jawControl*. Delete the type curves and generate new type to make a *jaw* label.

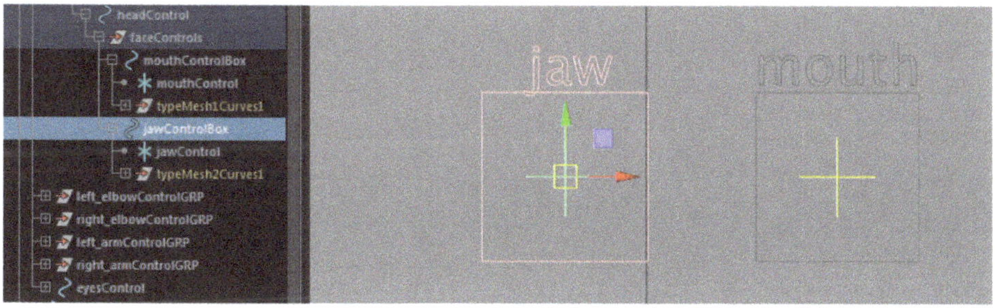

Set Driven Key with *jawControl* as the **Driver** and the *jaw* joint as **Driven**. In this case, we want the jaw to rotate in Z in mostly one direction—open—so key *jawControl.translateY* at its default **0** and *jaw.rotateZ* at its default **0.**

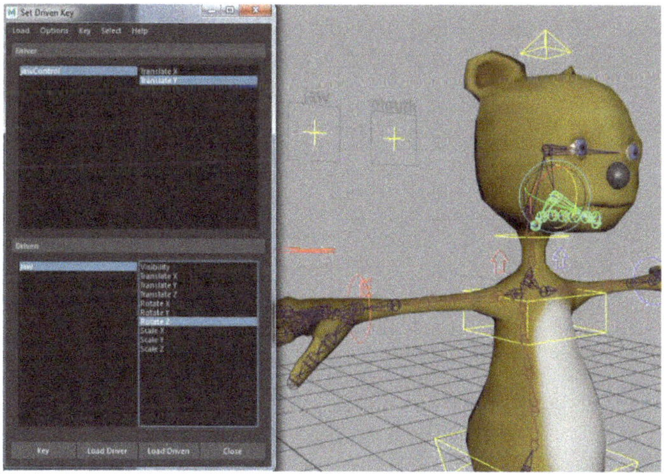

Then set a key with *jawControl.translateY* at its minimum (**-0.2**) and *jaw.rotateZ* at an extreme open position.

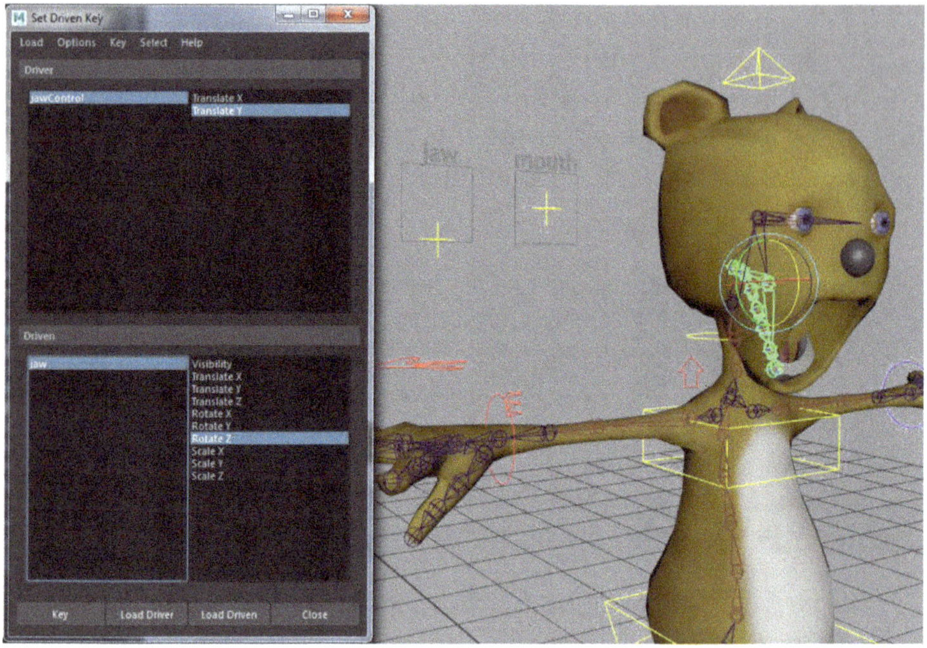

Now set a key with *jawControl.translateY* at its maximum (**0.2**) and *jaw.rotateZ* rotated positive in Z for an "extreme closed" position.

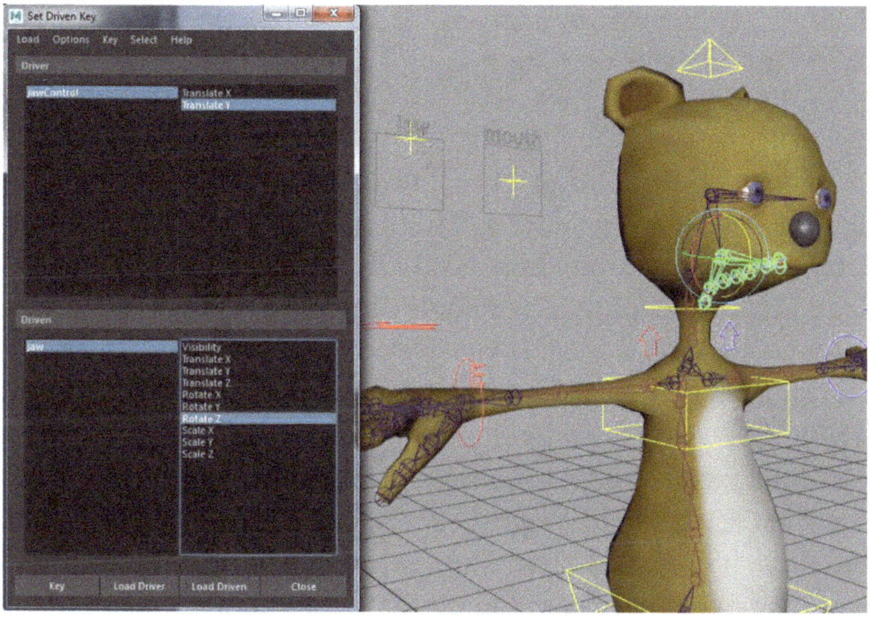

Now key *jawControl.translateX* and *jaw.rotateY* both at the default of **0**. Then set a key with *jawControl.translateX* at its maximum and *jaw.rotateY* at **45** or whatever looks like the extreme for your character. Then set a key for the other extreme.

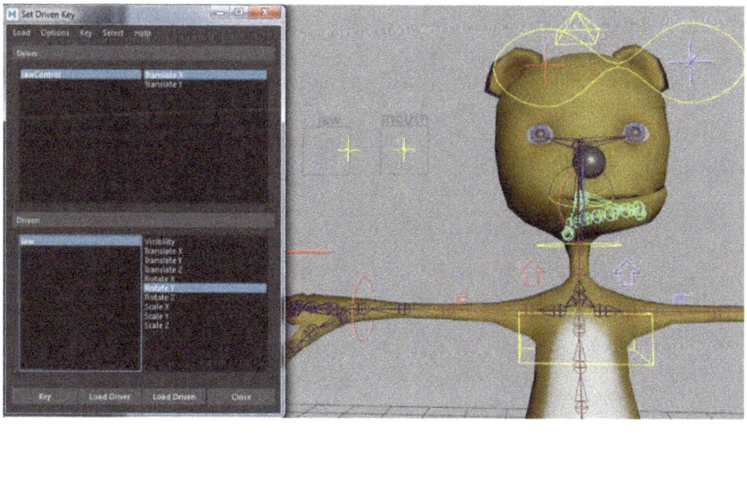

Together, these two controllers can drive the most common mouth shapes.

With this procedure, you can create the rest of the facial GUI in the same way. One for the tongue, and others to drive the other blend shapes. One interesting case study is Package Man, downloadable at *http://www.rigging101.com/*. His facial setup includes a slider that controls the rotation of the jawbone, linear sliders controlling the eyelids and eyebrows, and a slider controlling bend/squash blend shapes on a lattice deformer around the head for cartoon squash 'n' stretch. You can set up sliders to drive nonlinear deformers the same way (**Deform > Nonlinear > ...**). If you broke out your blend shapes into left and right shapes, you'll want to have separate sliders for the left and rights sides, or you can get fancy and use utility nodes such as the Condition node to create a box controller that slides between the left and right versions of a shape. In short, this same procedure can be used to drive just about any attribute value on a character rig! Here's a screen shot of the Package Man setup:

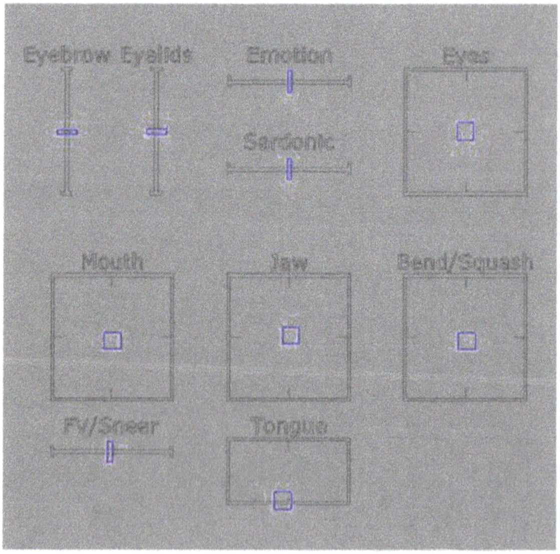

5.5 BEYOND THIS LESSON

When you are done setting up your facial GUI, make sure the whole thing is grouped and parented under the *headControl.* You can also add an attribute to the head control (or root control) that drives the visibility of the facial GUI group, so you can turn it on when doing facial animation, and turn it off otherwise to avoid screen clutter. Another convenience is to parent or constrain a camera to the facial GUI, so you have a dedicated camera for facial animation.

If you don't like the clutter of a facial GUI floating next to your character's head, you can create a controller with attributes for each of the face blend shapes using the Connection Editor. For example, create an attribute for each of the facial blends on the *headControl*, and use the Connection Editor to make a direct connection for each of them to the blend shape node. Whether you use a GUI or the Channel Box for facial animation is a personal preference up to you or your animators.

Yet another way to handle facial (and body) animation is with a Pose Library plug-in, such as the one found at *http://www.studiolibrary.com/.*

This lesson in facial setup represents the bare minimum set of controls for facial animation for a minor character with limited lip sync. Major characters usually have denser meshes and will require additional controls for nuance of expression. Start with a joint to control the nose (use the same techniques we used for the eye and jaw joints) as well as joints for the inner and outer brow of each eye and joints for the center and corners of the mouth. Beyond that, add more joints as necessary—realistic characters could have dozens of joints in a face rig. You'll find a layered approach that uses both blend shapes and joints will bring the variety of expression your character requires.

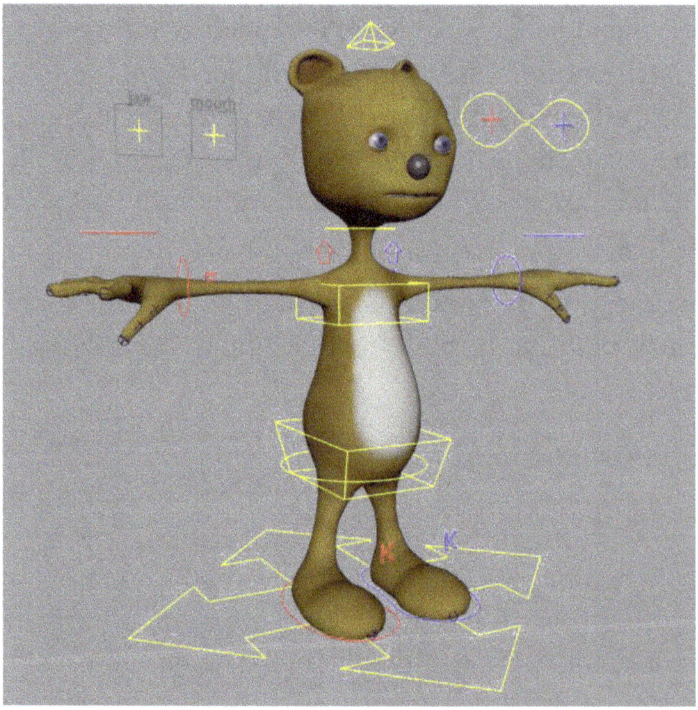

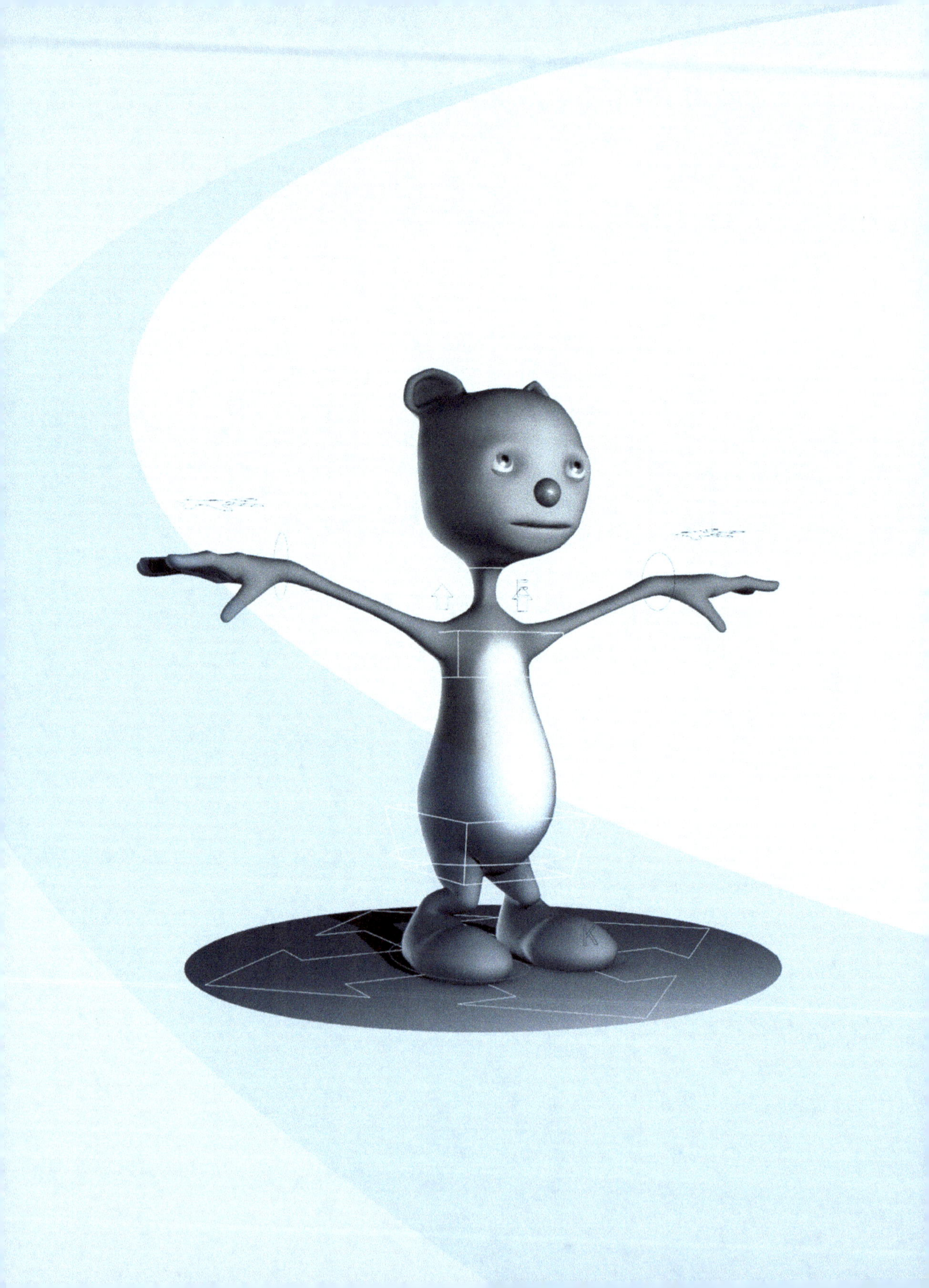

Part VI
Character Rig Testing

6.1 FINISHING TOUCHES

Before we enter the all-important rig testing phase, let's add a few final touches to make our animators and lighters happy.

First, let's add a smooth modifier to the rig. You've been able to preview with smooth mesh preview (toggling between the **1** and **3** keys on your keyboard with the mesh selected), but before handing this off to the animation department you'd want to lock the mesh off, making it unselectable. They'll find it difficult to unlock the mesh to smooth it every time they do a render, so we'll put a smooth modifier on the rig that they can access any time (and even keyframe!).

Before you do this, this is your last chance to delete non-deformer history on your mesh (delete all history except deformers; skin clusters and blend shapes are deformers, so they won't be deleted). So, save a copy of your rig (in case something essential gets deleted) and then **Edit** > **Delete by Type** > **Non-Deformer History**.

Next, **Mesh** > **Smooth**. You'll see a *polySmoothFace1* node attached to *bodyGEO* in the Channel Box. The mesh looks smoother because each quad was converted into four smoothed quads, essentially making our mesh four times heavier—now you understand why we were being so frugal with topology during the modeling stage. This version of the mesh is still light enough to update in real-time with most GPUs, so we can leave this as the default. Notice, though, that the *polySmoothFace* node has a *Divisions* attribute to divide it yet again, so we can use *Divisions=2* as our ideal render resolution. Our control should allow us to toggle between these three levels of division.

Select the *masterControl* (animators first look for attributes that globally affect the rig on the *masterControl;* you can also create attributes here that turn on and off visibility of parts for changes of costume, hairstyle, texture, etc.). **Modify > Add Attribute…** and name it ***Smooth***. **Date Type: Integer** with **Minimum 0, Maximum 2**, and **Default 1.**

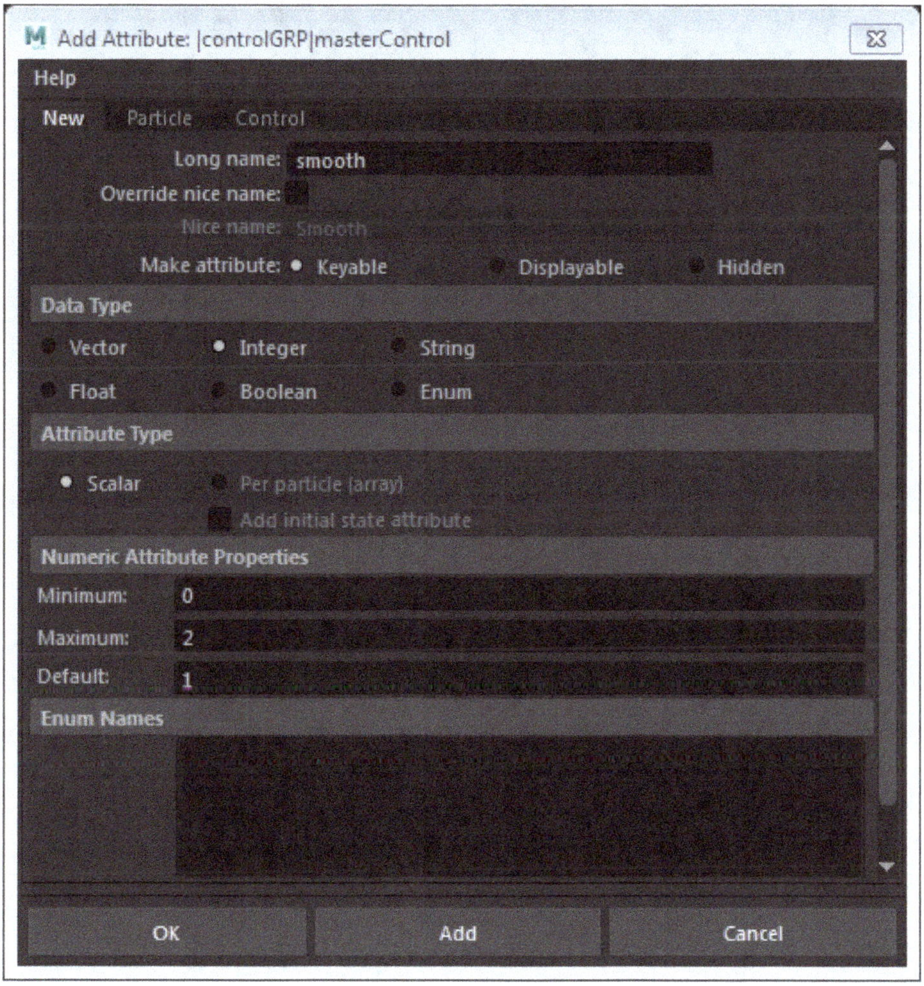

Windows > General Editors > Connection Editor. With *masterControl* loaded on the left, select *bodyGEO/polySmoothFace1* and **Reload Right**. Connect *masterControl*'s ***smooth*** attribute to *polySmoothFace*'s ***divisions*** attribute.

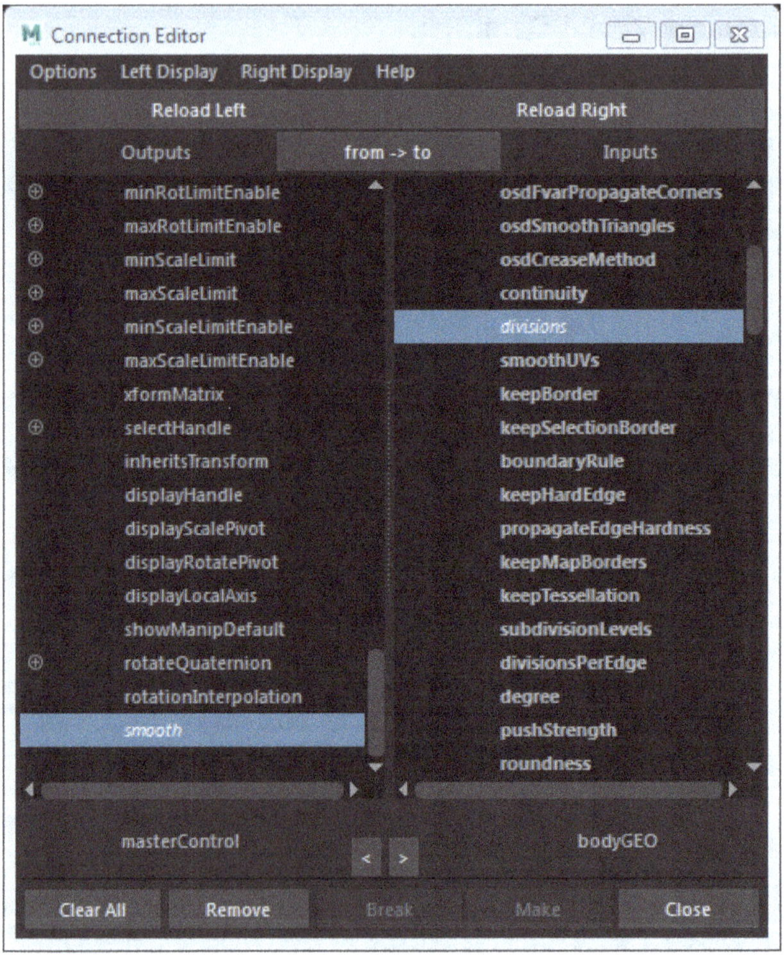

Now the resolution of your rig is selectable from the *masterControl*.

Another feature that will become essential once you start producing for a production pipeline is a scalable rig. Reason: Sometimes the default size of your world changes. For instance, when you start collaborating with another studio that uses different standards and conventions, you'll suddenly find your rig is tiny compared to a huge set, or vice versa. So the ability to scale the whole rig becomes essential.

Select the *masterControl* and unlock its scale attributes we previously locked (use the Attribute Editor and right-click each scale value to **Unlock Attribute**). Test uniform scaling your rig by using the *masterControl.* Some things scale, some do not. In general, things in a hierarchy will scale (unless locked by a scale constraint) but things parent-constrained into a hierarchy will not. So the process here is to track down and document everything that is not scaling because they are neither skinned nor in a hierarchy below the *masterControl* object. You may have deviated by now, but in my rig I need to scale the *masterControl*, *skelGRP* (top of skeleton hierarchy), teeth and eyelid geo (since they are constrained but not skinned or parented), and *handControls* (also constrained but not parented). Selecting these together and scaling produces acceptable results. So we'll connect the *Scale* attribute on the *masterControl* to the *Scale* on each of these nodes.

TIP

Certain other things can break scaling, like clusters in a spine setup that are *relative.* You'll get good at troubleshooting these things as you work through this process!

Select the *masterControl* and load the Connection Editor. It should be already loaded on the left, so select *skelGRP* and **Reload Right**. Connect the ***scale*** attribute of each.

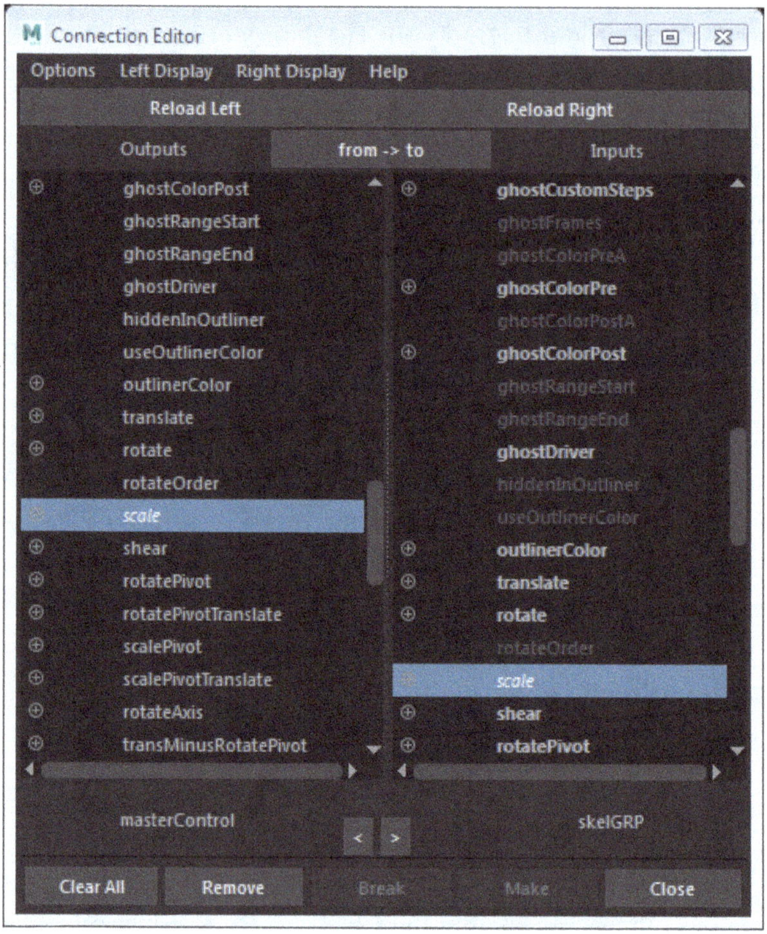

Reload Right for each of the items on your scaling list and connect their scale attributes to the *masterControl* scale. You could also add a Scale Constraint between these objects if you don't like the heavy-handedness of the direct connections. Test the scaling of your rig.

6.2 FILE CLEAN UP

Make sure your Outliner is clean and organized into no more than four top-level hierarchies: *geoGRP* (meshes), *controlGRP* (animation control rig), *skelGRP* (joints), and *miscGRP* (IK handles and other).

You should be able to **File > Optimize Scene Size**, but save a version first, as this action is not undoable.

Make sure that all nodes are named according to convention (including render/texture nodes!) and that display layers are organized and intuitively named.

Now that we have a reasonably usable rig, it is time to revisit each of the animation control objects we've made and make sure all unusable attributes are locked and hidden. Also check to make sure all controls are zeroed-out; animators rely on selecting all controls and returning their values to 0 to return the rig to a default pose; if any keyframable attributes are not zero at the default pose, freeze transforms and/or re-rig them.

6.3 CALISTHENICS TEST

It is time to start the important and iterative phase of rig testing. It is best if you have someone else test your rig for you during this phase, as a second pair of eyes can help identify problems. Sometimes problems are actual bugs in the rig, for which you have to troubleshoot and repair; sometimes problems are User Interface issues, where controls aren't as intuitive or easy-to-use as they could be. No two rigs are exactly alike, so this is a creative endeavor; finding ways to simplify and improve a rig requires innovative and outside-the-box thinking.

A useful exercise for starting the rig testing phase is to put it through a basic calisthenics test. The idea here is that we will test each control by animating it to its extremes—the entire recommended range of motion for each degree of freedom on each joint.

TIP

The recommended range of motion may differ from the actual range of motion depending on how you limit your rig. Some animators prefer a rig limited to the recommended range of motion so it's unbreakable, while other animators prefer a fully *unlocked* rig where controls allow you to push the rig *off-model* or to extremes that are not anatomically possible, so they have the freedom to selectively break the rig during *smear frames,* an animation gimmick where you push an extreme pose for one or two frames to heighten impact.

Before we start animating it is helpful to create a *Select All* icon on the shelf. Drag-select all animation controls (if your display layers are set up correctly, only the controls should be selectable because skinned geometry is on a reference layer). Click the gear icon to the left of your shelves and go to **New Shelf**. Name it ***Generikat*** (or your character's name). **Create > Sets > Quick Select Set…** name it ***Select All*** and click **Add to Shelf**.

TIP

You can use this method to create any number of selection sets to populate a shelf for ease of use when animating. It's also useful to have *Select All Face* and *Select All Body.*

Make sure your time slider is at frame 1. Hit **S** to set a key on all keyable attributes on this frame. Move to frame 10 and select *root-Control.* Translate it in Y as low as would ever be comfortably done in animation. Set another key. Then move your time slider back to frame one and with your middle mouse button click on frame 20 and set a key. Using the middle mouse button doesn't update the pose so you are essentially copying the default pose from frame 1 to frame 20.

Do this for every range of motion on every control, starting with the main controls and working toward the extremities in 10-frame increments. Don't worry about facial controls yet, we usually perform a separate facial test. Extend the time slider for as many frames as necessary to capture the entire range of motion in the body rig.

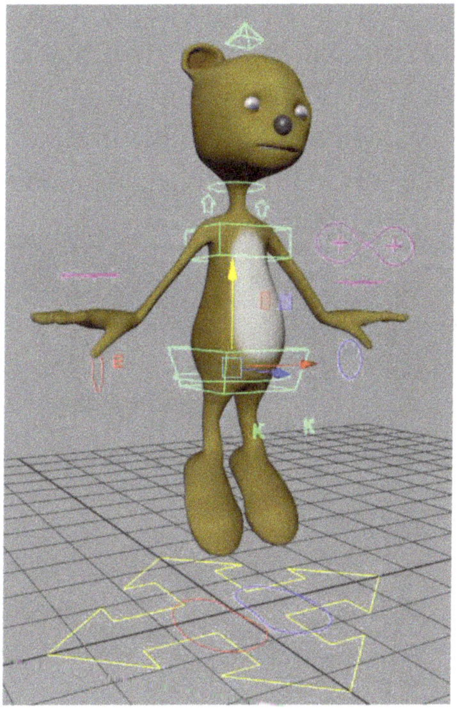

Feel free to make rig changes as you work through this process. For example, you may find you prefer your knee pole vector controllers to follow the orientation of the foot, so you may decide to parent constrain each *kneeControl* to its corresponding *footControl*. Make any modifications you think will make achieving good poses easier.

When you are done, center your character in view, rt-click the time slider and go to **Playblast … (options)**. Check **Save to file** and output a rendered test movie (A *playblast* is a screen capture, so only show layers you want to be visible in your rig test movie).

The purpose of this test is to demonstrate a working rig. Many animation directors will require such a test to approve a rig for use in production. One of the reasons to standardize a production rig—that is, use the same re-proportioned rig for every character—is so that you can save this calisthenics test and import it onto all future rigs during the skinning and rig testing phases. There are various ways to import and export animation curves to a whole similar hierarchy, including Maya's own ATOM exporter. Turn it on in the **Windows > Settings/Preferences > Plug-in Manager** under ***atomImportExport,*** and you will then see **ATOM** under the **File** menu.

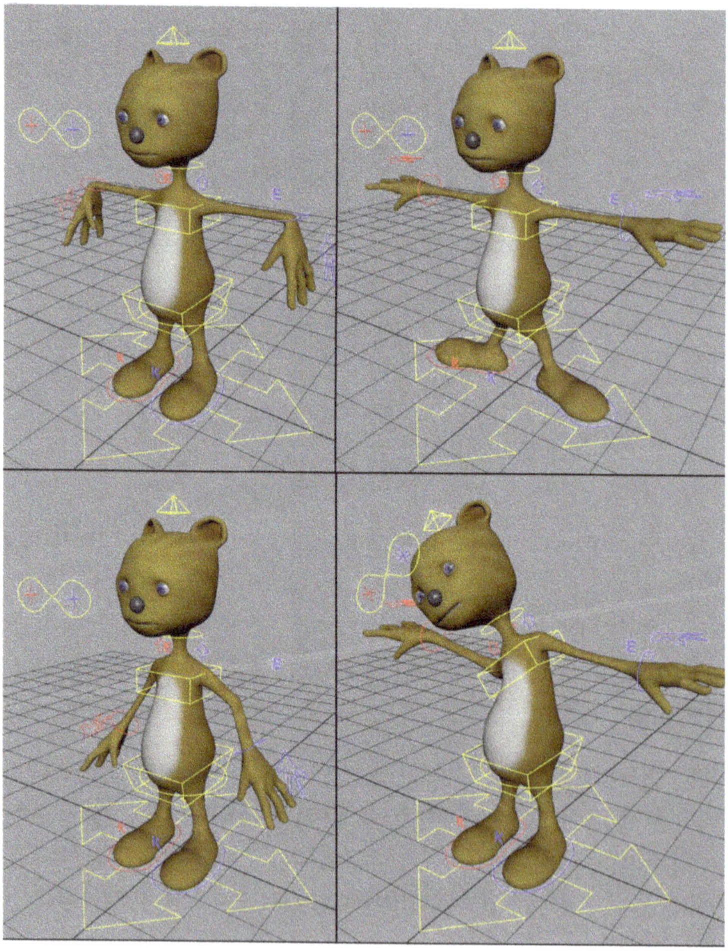

6.4 LIP SYNC TEST

The best way to test a facial rig is to do a short lip sync exercise. For this we'll need an audio file. There are dozens of web sites offering audio clips from film or TV shows, or you can record your own clip with the mic on your phone or computer. Keep it less than ten seconds and make sure there's a variety of phonetics in the clip (different vowel and consonant sounds). Acceptable audio formats include .wav or .aiff files. Start by placing the audio file in your Maya project directory under the *sound* subfolder. Then **File > Import...** and import the sound file. Now when you rt-click on the time slider and go to **Sound** you should see the name of your audio file. If the sound is loaded you'll see a waveform on your time slider. Note that **Playback Speed** must be set to **Real-time** for sound to be heard during playback.

To prepare for facial animation, you'll need to decide how you're going to keyframe blend shapes. The Shape Editor is one way, as each blend shape is exposed and keyable there. The drawback is it's another floating window to keep open, as well as the fact that blend shapes won't get keyed when you key all control objects in the viewport. A second way is to use the channel box, but that requires selecting geometry in the viewport to select the *faceBlends* node, which is usually a bad idea. The ideal way is to finish setting up a facial GUI, as we demonstrated with the mouth and jaw sliders. But this takes some time and lots of testing to get right. In a hurry, you can just add the rest of the blend shapes to a control object, as I've done here with *headControl*. Add an attribute for each blend shape with a minimum value of 0 and a maximum of 1. Then use the Connection Editor to connect each attribute to its corresponding blend shape.

Notice I created a separator above the blend shapes; I simply added an attribute of data type ***Enum*** with the long name ______ and the enum name ______ (six underscores).

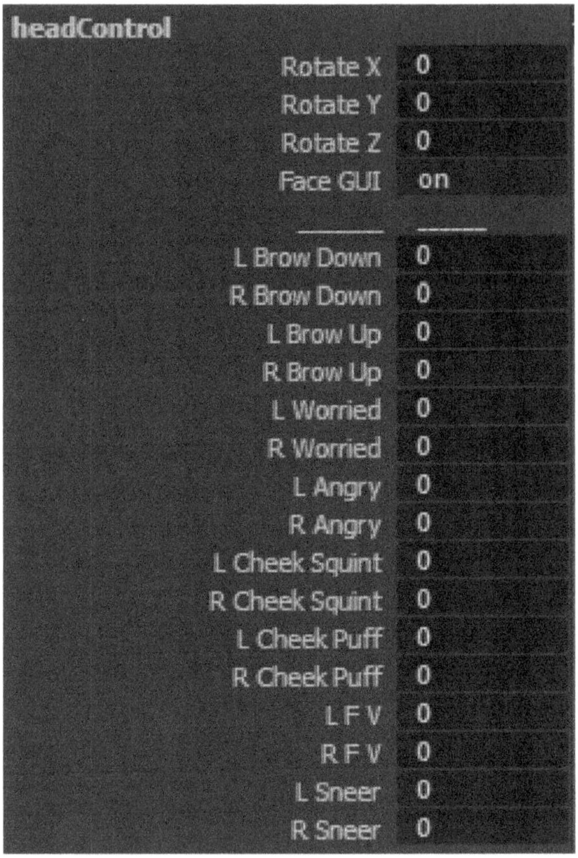

Another way I've prepped my scene is by adding a new perspective camera called ***faceCAM*** constrained to the *headControl* and locked, with a wide enough **Focal Length** to see my facial GUI. I've added a tongue control to the facial GUI that controls the curl of the tongue on the Y axis (all the tongue joints are driven in rotate Z by SDK) and the length of the tongue on the controller's X axis (Scale X on all tongue joints). The ability to curl and extend the tongue is essential for "l" and "th" phonemes.

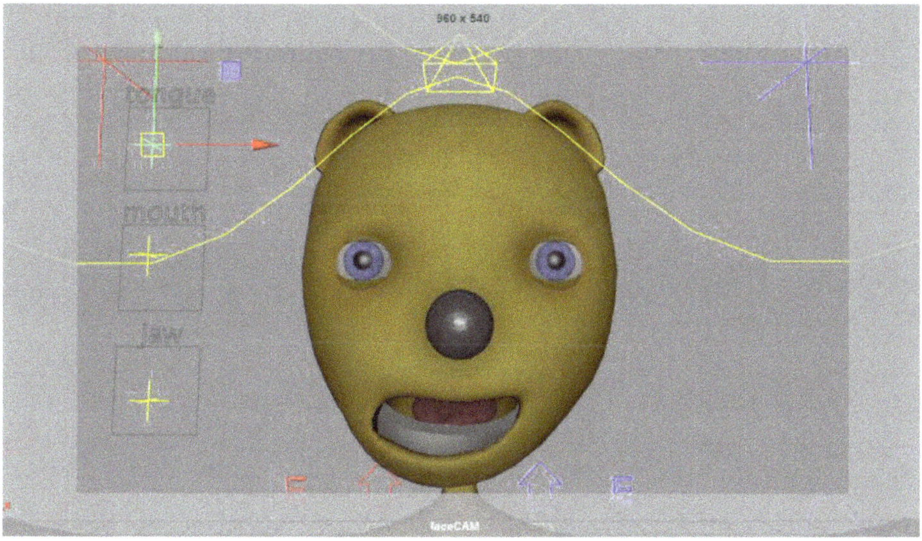

So now we're ready to set keys for our lip sync facial animation test. We'll be setting keyframes in two passes: on the first pass, we'll concentrate on wide/narrow and tall/short mouth shapes, so we'll only use the facial GUI. On the second pass, we'll add the rest of the visemes and full facial expression using the rest of the blend shapes on the *headControl.*

TIP

Phonemes are distinct units of sound in speech. Visemes are the distinct shapes a mouth makes during speech. Several phonemes can correspond to one viseme, as some sounds look the same on the face when produced. Though there are around 44 phonemes in English, we have far fewer blend shapes representing visemes.

The best way to break down lip sync is to first do a rough pass in which you concentrate on keying the jaw open and close on the accents. You can place your hand on your chin and say the phrase, taking note of when your mouth opens most widely. These are the accents in the phrase. Key the jaw opening and closing for each accent, and then think about wide and narrow shapes; "oh" and "oo" sounds are narrow and closed; long "i" sounds are narrow and open; "aay" and "eh" sounds are slightly wider and "ee" sounds are widest. To finish this rough pass, make sure you add closed-mouth shapes for every "b," "p," and "m" stop.

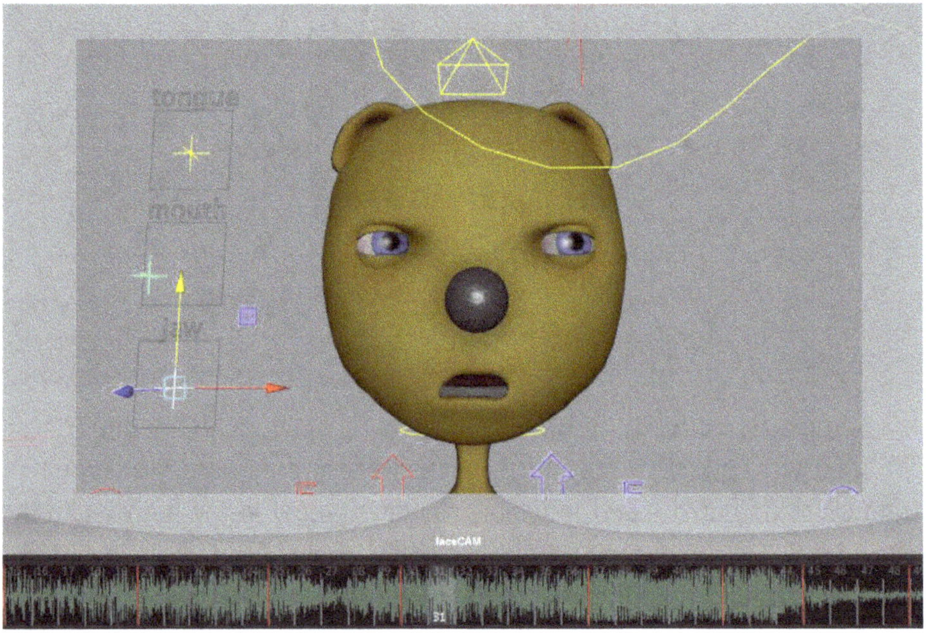

TIP

In the Time Slider Preferences (icon to the far right of the range slider) you can adjust the height of the time slider to better see the peaks of the waveform.

On the second pass, add the fricatives ("f," "v," "th" shapes) and dial-in attitude and expression. You'll find that every syllable does not need its own keyframe; in speech, we have the tendency to "blend over" certain sounds. This is why we concentrated solely on open and closed shapes during the first pass; these are usually the necessary ones, and other phonemes and some lost syllables tend to blend and merge together. Another trick is to let keyframes precede each sound by a couple of frames; we see mouth shapes before we hear the sound. If you don't offset the lip sync in this way, it will appear to lag behind the sound, even if you are hitting every syllable exactly. Also, use a mirror or phone camera for reference, and don't be afraid to exaggerate!

When you are done lip syncing, playblast and evaluate. This is the time to add or edit blend shapes, or to add additional facial controls if your performance is "flat."

TIP

The first lip sync test will reveal the deficiencies in many blend shapes. Open the Shape Editor, select the offending target, and **Shapes > Rebuild Target**. You can adjust the target shape on the fly then return to animating.

6.5 FINAL PUBLISH FOR ANIMATION

Now that you've tested and iterated on the rig, it's time to publish it for use in your first animation scene. Because you can anticipate more rig adjustments in the future (the process of rigging is never really done), standard practice is to reference your rig into a scene file rather than animate in the rig file itself. With file referencing, one Maya scene file—the *rig file*—is referenced into another Maya scene file—the *animation* or *shot file*. Changes made to the rig file are updated in the shot file every time the shot file is reloaded. In this way, you can fix errors and make enhancements to the rig while you animate, and those changes will be reflected in any and all shots that the rig is referenced into.

TIP

A *shot* is the elementary unit of an animated production. A shot is usually one camera move or take between cuts. Multiple shots make up a *sequence* (a series of related shots) and a series of sequences make up an *act* or *show*. Use these units of production management to help organize your Maya scene files. Note that individual shots usually, but not always, equate to individual Maya scene files, thus these are referred to as *shot files*. Shot files contain referenced rigs, and usually also referenced *props* and *sets* and sometimes *lighting rigs* and *camera rigs*. Animation, simulation/effects, and lighting are usually done locally in the assembled shot file.

If you haven't caught on by now, the key to rigging is organization and tidiness. Make sure once again all controls are zeroed-out and display layers are set up so that skeletons and other junk are invisible and only controls can be selected.

Make sure any test animation you may have done is deleted. You can **Edit > Delete by Type > Channels** (by default SDK is left intact) or with all controls selected you can double-click the time slider to highlight all keys, rt-click and **Delete**.

When your rig scene file is ready, save it and open a new scene. Using **File > Create Reference…** bring in your rig scene file as a reference. Note that in the options you would turn on **Group** to put the file's contents into a group node to keep things tidy; optionally, you could create a top-level group node in your rig file with the name of your character.

References can be manipulated in the **File > Reference Editor**. They can be loaded and unloaded (handy when working on a scene with many heavy characters), reloaded (handy when rig changes are being made on the fly), duplicated, and deleted.

Any change (including animation) made to a rig in the shot file is local to the shot file; they do not affect the original rig. So an animator might use a deformer or constraint in the course of animation, but these extra nodes would not propagate to other shots. If the animator decides a certain deformer or constraint would be useful in multiple shots, it would be better to add it to the rig file.

TIP

Maya has a couple of optional schemes to further help you control how animators interact with your rig, and what nodes they have access to.

Character Sets are one solution (under the **Key** menu). Creating a Character Set allows you to gather all attributes from different nodes into a character set for ease of access. It creates a slightly different workflow for the animators, though; it enforces a pose-to-pose approach rather than animating in passes which is more common in CG character animation. Since animators tend to be stuck in their ways, this feature has limited use in the industry. Character Sets do offer flexibility in importing/exporting and retargeting animation, though. Feel free to look into Character Sets if you want to take advantage of this feature in your character pipeline.

Another publishing scheme in Maya is its Assets system. This is useful in larger, more complex pipelines where you want to enforce workflow standards through the use of templated scenes. Used with file referencing, Assets are a good way to modularize an entire production pipeline. They are beyond the scope of this book, but you can dabble with it if you select a group of objects and **Create > Asset > Create Asset with Transform**. Think of assets as fancy group nodes.

Now that you have a referenced character in a shot file, import sound clips, import sets and props, create a shot camera, and start animating!

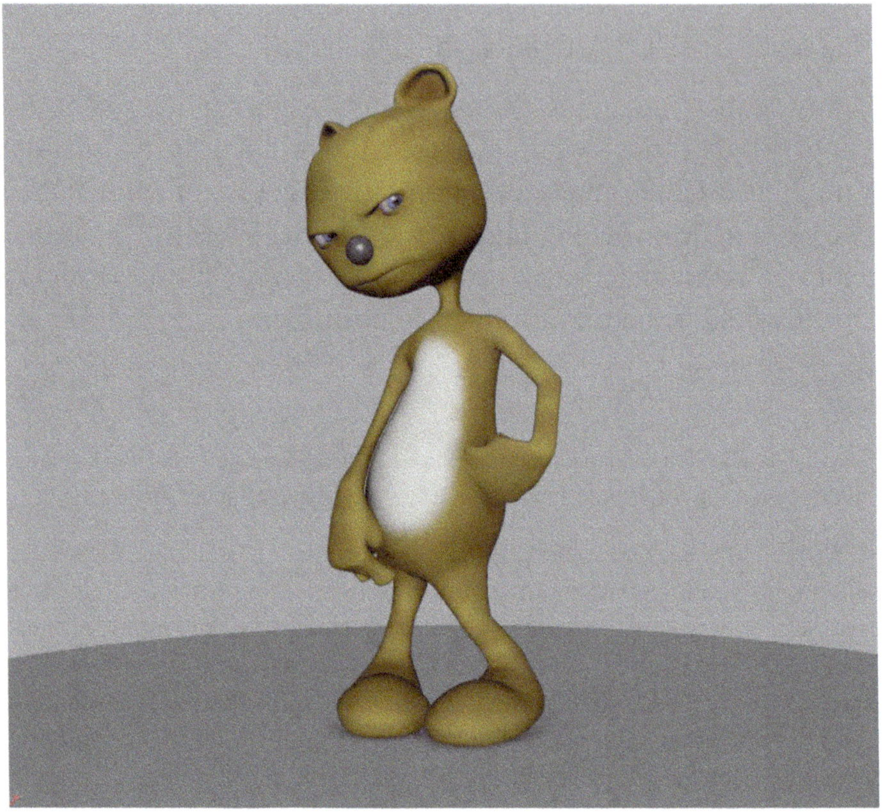

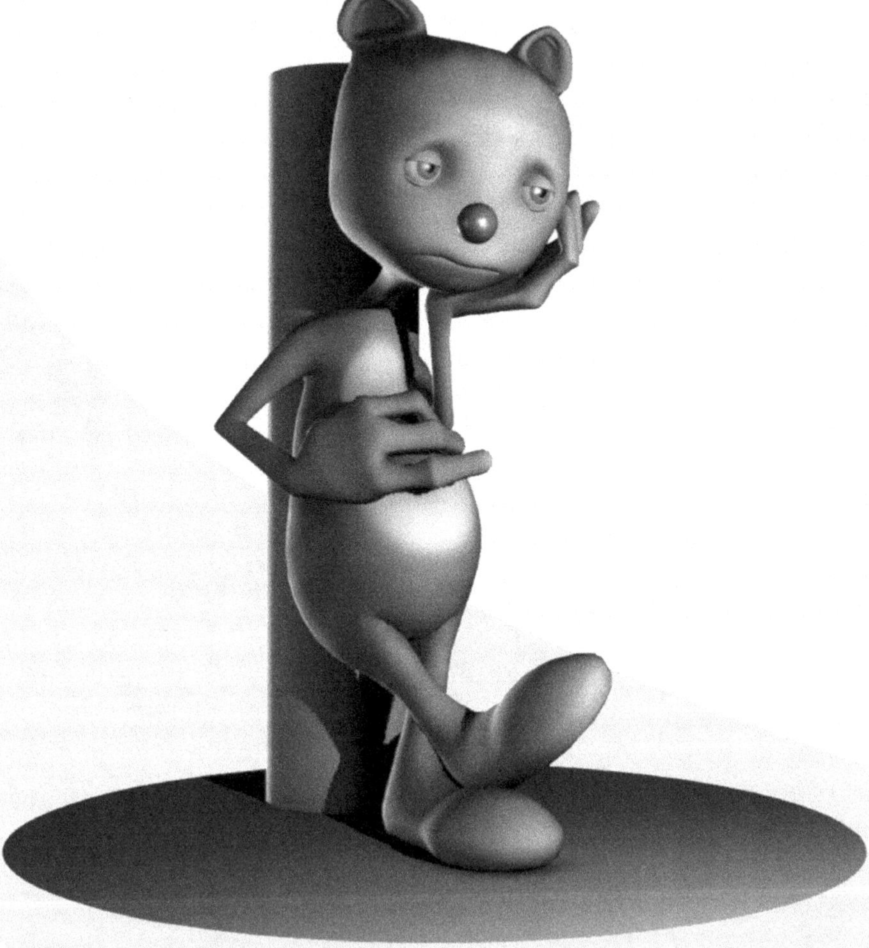

PART VII

ANIMATION

Your rig is tested and bulletproof. You're ready to animate. What now?

Here is a brief rundown of the stages of animating a shot.

7.1 STORYBOARDING AND ANIMATIC

The first step to animating a shot or sequence is to storyboard it. The key to storyboarding is speed and flexibility. You are experimenting with story ideas through the relatively fast-and-dirty medium of simple line drawings. Especially when storyboarding for yourself or a small team, do not worry about line quality or expert rendering; a lot can be conveyed by simplified figures and abstracted sets—sometimes little more than horizon lines. Remember the purpose of a storyboard is to work out story *beats*—the atomic units of story—and story *pacing*—or how fast the action unfolds as told through beats that set-up, beats that build anticipation and beats that pay-off. Since this is an iterative process, with the need to discard beats, add beats, and rearrange beats, don't draw your initial storyboard like a comic strip on one piece of paper! Use note cards, or separate files/layers if working digital. Individual storyboard drawings are sometimes called *sketches.*

How many sketches do you need for a storyboard? Well, one for every action and/or beat, at minimum. For complex actions, you may need before and after sketches (called *A/B sketches*) or beginning/middle/end sketches to successfully describe the action.

Once you have all of your story sketches assembled, you can use simple video editing software to output an *animatic,* which is your storyboard set to timing. Often there will be the addition of a *scratch track*—temporary score, sound effects, and dialog—to help specify timing. The purpose of the animatic is to determine *timing* and *editing*—the exact number of frames needed for each action to resolve on the screen. It is important that this is determined before a CG production can begin, because you will set your individual Maya scene files to the number of frames based on each shot's duration in this animatic.

Once an animatic is "locked" —that is, edited so that timing is worked out to an exact frame count—you can develop a *shot list* and an *element list.* A shot list contains *dope sheet* information (that is, frame count and dialog/beats for every shot) as well as a list of assets needed for the shot. From this is derived an element list (characters, sets, props, and effects) and *element production* (sometimes called *asset production*) can begin.

7.2 LAYOUT

All elements or assets are assembled through file referencing. These include characters, sets and props. A camera rig is imported or referenced in, or a shot camera is created.

TIP

Setting up a Shot Cam. **Panels > Perspective > New.** Name it ***shotCAM***. In the Attribute Editor, set **Focal Length** to 45 or 50 to approximate the lens of the human eye (the default 35mm is decidedly wide angle to accommodate modeling). Under **Display Options** turn on **Display Resolution** and **Display Gate Mask,** with 1.100 **Overscan.** Once you get something you like hit the **Presets*** button and **Save Camera Preset** to use it in all your shots. Resolution is set in the Render Settings window. Make sure *shotCAM* is the only camera listed under **Renderable Cameras** and set the **Image Size** preset to HD_540 for test renders. For final render, you can bump this to HD_1080.

The time slider range is set to the length of the shot as determined by the aforementioned *shot list*. The scratch track can be brought in to provide the sound queues to animate to. Use the **Offset** attribute in the sound node's Attribute Editor to slip the scratch track to the right starting frame for the shot (if this isn't the first shot in the sequence).

If this is part of a sequence, you'd first create a set and a rough lighting rig, and at this point you'd bring those in through the Reference Editor. If this is a standalone shot, you may import props and build the set here. You can always export it later for use in other scenes (**File > Export Selection (options) > Keep only a reference.**)

You can also set up rough lighting at this stage and use **Lighting > Use All Lights (7)** to animate to. Mood lighting sometimes affects performance, so place lights and shadows where they belong prior to animating. Don't worry about getting it perfect; final lighting will be done as part of the lighting/rendering/compositing phase, later.

Animate the camera, if there's a camera move. For a static shot, lock off the camera **(View > Lock Camera).**

The final task of the layout phase is *staging*. Move and/or animate the characters and props to the part of the frame they should roughly occupy; use the storyboard as your guide. Do not yet employ full character animation; simply keyframe the root node of the characters around just to keep them in or out of frame to pre-vis the shot.

Once the camera is animated and locked off, and staging of sets/props/characters is done, layout is final.

7.3 ANIMATION: BLOCKING

Animation is generally done in three passes. The first of these is the *blocking* phase, a term from theater, meaning a rehearsal where actors practice hitting their marks on the stage without the distraction of acting. This is a bit different conceptually, as we already took care of "staging" during the layout phase and now we are concentrating on hitting the key poses of our animation. As such, this is an evolution of what Frank and Ollie call *pose-to-pose animation;* that is, starting with the key poses and then doing breakdowns and in-betweens later. The effort here is to establish overall timing and defining strong poses as our key poses, or *keys* (yes, these are also referred to as keys—we call lots of things "keys" in CGI just to confuse you). Key poses are the storytelling poses; the ones that can't *not* be there or a story beat would be lost. Start the blocking pass by defining these key poses.

The first step is to gather reference and plan. The best reference is to get up and act it out on camera. Knowing you'll exaggerate even more during animation, act it out as best you can in front of a mirror, then in front of a camera, and bring that footage into the computer for frame-by-frame reference. Get ideas from videos you find online. As you gather reference, start the planning phase by *thumbnailing.* Thumbnailing is the process of drawing quick, gestural sketches to define the key poses, so called because these drawings need be no bigger than your thumbnail, since they are usually nothing more than lines of action with some body masses roughed-in to define the general pose. You can thumbnail while roughing-in timing by using the grease pencil tool to draw your thumbnails on specific frames of animation **(Panel Menu: View > Camera Tools > Grease Pencil Tool)**. The grease pencil tool creates its own keyframes on the time slider that you can move to adjust timing just like any other keyframe (by shift-clicking the key tick and dragging it on the time slider). Draw all your key poses and then play forwards or Playblast to evaluate timing. Move the keys around until the overall timing feels good.

Once you've established overall timing, it's time to fill in a few more frames. The key poses were the storytelling poses; next we'll do the extreme poses, or *extremes.* The extremes define the beginning and end of every action. For example, the storytelling key pose of a baseball pitcher pitching a ball might be the moment she's letting the ball go. The extremes of this movement would be the wind-up before the key pose and the overshoot frame after the key pose, when all the arcs and trajectories of motion are at their extreme limits. Going further, we could define the relaxed pose before the wind-up and the settle after the overshoot. Do this for every key, finding the extreme range of motion and placing your extremes around every action.

Finally, fill in your *breakdowns.* Your breakdowns are the poses midway between your extremes that define the height of your arcs (everything moves in an arc in animation). Though the breakdown pose is spatially halfway between the extremes, place each one on the timeline favoring one extreme or the other; that is, closer on the time slider to one extreme. In this way, you are defining the *slow-in/ slow-out* (or *ease)* of each action, or the gradual build-up and/or loss of momentum through the action. If these concepts are foreign to you, run through a bouncing ball animation tutorial; there are plenty on YouTube.

Now that you've worked it all out with the grease pencil, blocking out the scene will be a breeze. First, rt-click the Time Slider and toggle the **Enable Stepped Preview** option. This will visually hold each key pose until the next one, allowing your animation to play back like a flipbook. In this way, you can begin to block out a pose with your character rig for every frame you've planned with the grease pencil, and upon playback it will hold the poses without any distracting incorrect auto-tweening.

TIP

In the old days (like two years ago), people referred to the *stepped* phase of animation, and then the spline phase or *splining*. They were referring to manually changing all the tangents of the animation curves to *Stepped* for blocking and then changing them all to *Spline* or *Auto* tangents to begin refining. Thanks to Stepped Preview, it's now a little easier to toggle between these states.

7.4 ANIMATION: REFINING

The central phase of animation is what I call the *refining* stage, but others may call the *splining* or *smoothing* stage. This is the stage that you turn off stepped preview and begin focusing on the *in-between* frames, or the *tweens.* That is, every frame of animation that falls in-between the keyframes that you've already set for the storytelling poses, extremes, and breakdowns. In computer graphics, tweens are calculated automatically by interpolating the remaining frames based on the tangents in and out of each neighboring key. Bad CG animation is made when this default interpolation is left alone; the resulting actions seem lifeless and swimmy, and lack snap. Good CGI comes from the knowledgeable manipulation of tweens. There are two primary methods used to tween, and both can make good CG animation, but only one is the correct way.

The first method of tweening is the brute force method. That is, set a key on every attribute on every frame. Though this is the process most familiar to traditional animators, this is the wrong method to use in CG pipelines.

The correct method of tweening is to avoid setting keyframes on your tweens, but to skillfully manipulate tangents and curves in the graph editor to bring life to your animation. From this stage onward, most of your animation will be done in the Graph Editor **(Windows > Animation Editors > Graph Editor)** using your camera view mostly for reference/playback. Some animators will turn off the visibility of their rig controls at this point since they will mostly be moving keys around in the graph editor from here forward. The reason to use tangents rather than keying every frame is twofold. For one, with practice it's easier to achieve smooth results, free from the jitter that comes with frame-by-frame animation. The second reason has to do with iteration; you can change the nature of the performance easier with sparse, organized keys—by scaling/stretching them, flipping them, copying and pasting them, or overriding them with keys on new animation layers. In short, you can achieve an "art directable" performance, able to incorporate notes easily without redoing everything.

Another concern during the refining stage is that of offsetting keyframes and adding overshoot, so that not every body part "hits" noticeably on the same key frames, and instead there is follow through and overlapping action.

The result of the refining stage is that every existing keyframe interpolates smoothly and intentionally to the next, and the overall performance is smooth, organic, and interconnected. Everything is constantly in motion, but at any given time only one thing commands our attention, and every secondary motion reinforces the primary action. Check that all the arcs of movement are smooth, and that poses and silhouettes are staged to maximize readability.

Some animators prefer not to start facial/lip sync animation during this phase, since it slows momentum, and instead concentrate on such details during the next phase: polishing.

7.5 ANIMATION: POLISHING

During the polish phase you usually aren't changing any major poses, but instead you're adding the details that breathe life into a performance. Facial and finger animation are fleshed out here, and secondary or overlapping action is added (things like hair bouncing, clothing flapping, etc.). Pay special attention to *contact frames*—frames of animation wherein any part of the character is making contact with another object (for example, a hand holding a ball should have every finger resting believably on the ball in every frame). Analyze every point of contact frame by frame to fix interpenetration and to make the contact look believable.

Also during polish you can add small imperfections to make your performance less "computery"; a small mouth twitch, a micro-expression representing a passing thought. Blinks and eye-darts (small sudden movements of the eye) will add to the illusion of life. As a rule of thumb, animated characters blink every time they have a new thought or feeling.

7.6 LIGHTING AND RENDERING

Once the animation performance is done, you'll begin to work on final lighting and render setup. This is a subject that can fill many books, but an independent animator must know enough about the subject to make presentable showreel footage or animated shorts. Good lighting is essential to showcase and dramatize the action; bad lighting can make your animation look bad. Familiarize yourself with the vocabulary of direct and indirect light sources and how they react with various surfaces. Lighting is a process that involves setting up lights and environmental effects like fog or particulate matter. Effects such as dynamics, cloth, hair, fur, and fluids are usually introduced into shots at this stage as well. Lighters also tweak shaders and textures during this process, and they set up render layers and passes/AOVs so that they can further manipulate the final look in a compositor like Nuke or After Effects.

If you are new to this, it is essential you get to know the basics of the **Render Settings** window first; here you can control output formats and resolutions and set all the parameters in your renderer of choice. For simple one-scene shorts you may find everything you need in Maya's simplest renderer, **Maya Hardware 2.0**—it's merely the viewport renderer! The advantage here is what-you-see-is-what-you-get in terms of lighting; what it looks like in the viewport is what it will look like when output. For anything more advanced, though, you'll need a standard renderer like Arnold, Mental Ray, or Renderman.

The other two windows you'll spend a lot of time in during this stage is the **Hypershade,** which is the command center for render nodes (materials and lights, etc.), and **Render Setup,** which separates your scene into render layers and passes so you can control different aspects of the final render in a compositor. Both of these windows are under **Windows > Rendering Editors.** Get to know them well.

For best practices, render to sequences of frames (not video formats), preferably formats with alpha channels for transparency and lossless compression (such as TIF, PNG, PSD, TGA). Never use lossy file formats like JPG or GIF unless it's a preliminary test render. Also understand Maya's linear color space and how to embed an output transform if using anything other than a 16- or 32-bit file format like

EXR or TIF. I recommend using the default of EXR so that you don't have to worry about output transforms and you can take advantage of a linear workflow. The only trick is learning how to work with EXR in your compositing package.

Eventually you'll learn what's easier to do in a compositor than in 3D; stuff like motion blur, depth of field, screen-space effects like film grain and noise, or any kind of blur or color correction effect.

Yes, it's a lot to learn, but start with what you need to know—what's needed to tell your story—and work from there.

POSTSCRIPT

Now we've followed the development cycle of Generikat from a line drawing to a fully rigged, animated character. This set of tutorials was meant to demonstrate the most common practices in 3D character development in the simplest possible scenario. Much more complex characters can be set up by utilizing the same procedures demonstrated here.

Once you understand these fundamentals, it is time to shop around for an automatic rigging system, or an *auto-rigger*. Maya comes with the Human IK system which has good compatibility with game engines, but lacks the features most cinematic animators require. There are popular commercial systems like Rapid Rig or The Setup Machine, as well as free alternatives like Advanced Skeleton and mGear. The best systems will be modular/extensible and have stretchy/bendy limbs, IK/FK matching, and space switching (e.g., hand IK switches parents to follow different parts of the rig, so it can either move with the upper body or stay locked in space while the upper body moves). Additional tools may include a picker GUI (a floating window to pick parts so animation controls can be hidden in the viewport), a pose library (GUI for saving and importing poses and animation clips), and pose mirroring/flipping.

Auto-riggers are essential to provide consistency to an animation pipeline—by standardizing the rig you allow scripts to access them by name or by hierarchy, as well as enforcing standards for rigging and animation. But only those with an understanding of the fundamentals can successfully set up or manipulate an auto-rigging system. Congratulations! After this set of exercises, you can now do this! At this point, you should feel emboldened to rig any of your own original characters, and create your own fully animated shorts. Choose an auto-rigging system, and customize it as you see fit. Most importantly, animate! The best riggers are animators, so practice animation, if only to perfect your rigging chops. But the best animators are also riggers, as they know how to get the best out of a rig. Most importantly, they never feel limited by their rig. Happy animating!

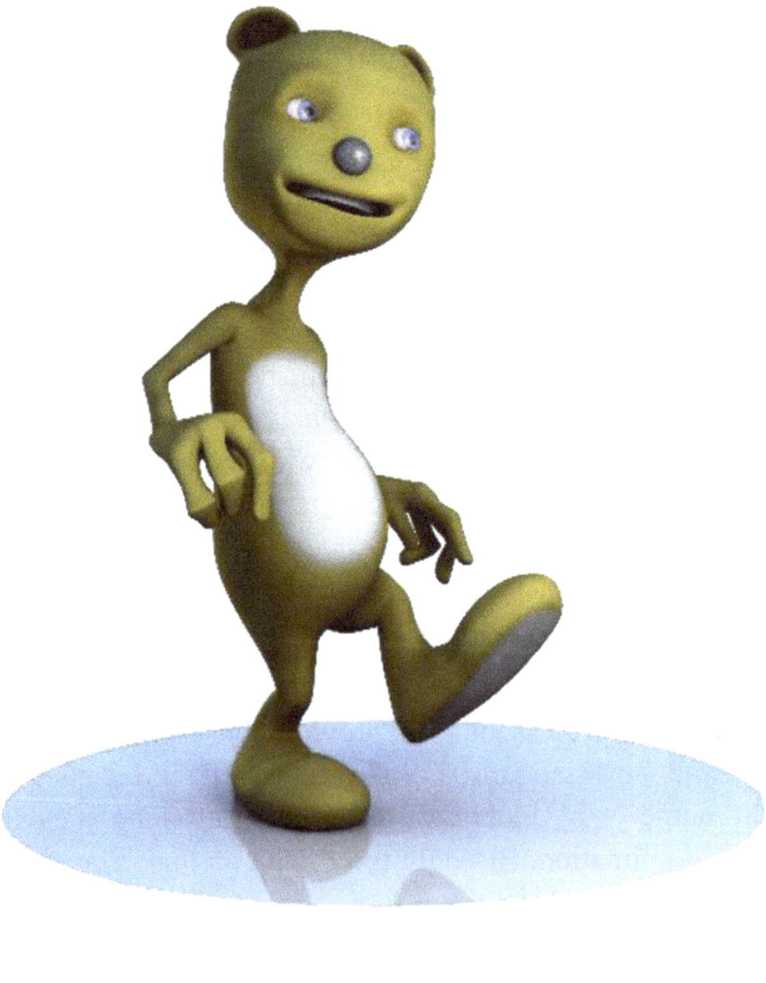

Index

www.ingramcontent.com/pod-product-compliance
Lightning Source LLC
LaVergne TN
LVHW080042210926
848481LV00011B/3258
9781683921707